Treatment and Disposal of Solid and Hazardous Wastes

Debashish Sengupta • Brajesh K. Dubey
Sudha Goel
Editors

Treatment and Disposal of Solid and Hazardous Wastes

Springer

CP

Editors
Debashish Sengupta
Department of Geology and Geophysics
Indian Institute of Technology Kharagpur
Kharagpur, India

Brajesh K. Dubey
Department of Civil Engineering
Indian Institute of Technology Kharagpur
Kharagpur, India

Sudha Goel
Department of Civil Engineering
Indian Institute of Technology Kharagpur
Kharagpur, India

ISBN 978-3-030-29642-1 ISBN 978-3-030-29643-8 (eBook)
https://doi.org/10.1007/978-3-030-29643-8

Jointly published with Capital Publishing Company, New Delhi, India.

The print edition is not for sale in SAARC countries. Customers from SAARC countries—Afghanistan, Bangladesh, Bhutan, India, the Maldives, Nepal, Pakistan and Sri Lanka—please order the book from:

Capital Publishing Company, 7/28, Mahaveer Street, Ansari Road, Daryaganj, New Delhi 110 002, India.

ISBN of the Co-Publisher's edition: 978-93-81891-48-3

This Springer imprint is published by the registered company Springer Nature Switzerland AG
The registered company address is: Gewerbestrasse 11, 6330 Cham, Switzerland

Preface

The present book deals with state-of-the-art research studies and technological innovations in modelling trends and remediation of solid and hazardous waste. Case studies of some of the most relevant topics in this area are dealt with and include waste utilization potential, development of integrated solid waste management systems, and reuse and recycling of waste based on green chemistry principles. Other interesting and pertinent topics discussed are remote sensing, GIS and linear programming, impact of sandstone quarrying on quarry workers and local residents, biodegradation of floral waste, and construction of wetlands for treating landfill leachate. In addition, the occurrence of natural radioactivity at elevated levels in groundwater and the environmental implications of high radiation along beach placers have also been discussed. Adequate protection and remediation methodologies are obvious outcomes of the present studies and are essential in view of the large-scale environmental impacts on a growing population. In addition, enhanced interest in air quality index, extreme weather conditions, and possible linkages with global warming also gives an additional impetus to the present book.

The subject matter of the different chapters in the book has been written by experts with teaching and research experience in their respective fields. We hope that this book will be useful not only to practitioners of Environmental Science and Engineering but also to state and central environmental monitoring and regulatory agencies and will contribute to the framing of future regulations.

Kharagpur, India

Debashish Sengupta
Brajesh K. Dubey
Sudha Goel

Contents

About the Editors

Debashish Sengupta is currently Professor at the Department of Geology and Geophysics at the Indian Institute of Technology Kharagpur. From January 2013 to December 2015, he was the Head of the same department. After completing his Ph.D. at the Physical Research Laboratory, Ahmedabad, he joined the Department as a Faculty Member in 1990. He has been involved in extensive research on applications of nuclear geophysics in Earth Sciences and modelling of radiation hazard and remediation for more than 30 years. He has also collaborated with international research groups from the University of Sao Paulo, Brazil; University of Roma Tre, Italy; and USGS.

Brajesh K. Dubey is an Associate Professor at the Indian Institute of Technology Kharagpur. He has more than 17 years of research, teaching, training, and industrial outreach experience in the areas of integrated solid and hazardous waste management, sustainable engineering, and application of life cycle assessment techniques.

Sudha Goel is currently an Associate Professor at the Civil Engineering Department, Indian Institute of Technology Kharagpur. She completed her B.E. (1990) in Environmental Engineering at L.D. College of Engineering, Ahmedabad, and M.S.E (1992) and Ph.D. (1996) in Environmental Engineering at Johns Hopkins University in Baltimore, USA, and has been working on various aspects of environmental engineering since then.

Chapter 1
Biodegradation of Floral Waste Under Aerobic Conditions with Different Microbial Inocula and Aeration Methods

Ved Prakash Ranjan and Sudha Goel

1 Introduction

India is one of the largest flower producing countries in the world. According to estimates of the National Horticulture Board for 2015, 317,000 hectares across the country were used for floriculture, producing 2.3 million metric tons of flowers of which 0.5 million tonnes are cut flowers and 1.8 million tonnes are loose flowers. Flowers are used in cultural and religious ceremonies, and floral offerings are an integral part of daily worship for most people in the country regardless of their religion. They are also used in major events and functions like marriages, cultural and other activities as decoration and as bouquets and garlands. Due to religious beliefs, most people in India avoid throwing flowers and other items that are used for prayers in the garbage, and instead put them in plastic bags and throw them directly in water bodies (Yadav et al. 2015). Thus, there are no suitable modes for disposal of this small but significant organic fraction of municipal solid waste. Further, cities that are religious or pilgrimage centres produce very large amounts of floral waste on a daily basis. For instance, Banaras, one of the holiest cities in the country, has no policy for disposal of the tons of floral waste that come from its many temples. Each day, waste material weighing 3.5–4 tons is left behind in this city of temples. Degradation of floral waste is a very slow process as compared to kitchen waste degradation (Jadhav et al. 2013) microorganisms do not live isolated, they coexist with many different microorganisms establishing relationships with each other which help them to degrade highly complex organic compound to simple stable end products. Therefore in the present work instead of individual isolates an attempt is made to develop microbial consortium for degradation of flower waste. Soil samples were collected from the area near the temple. Serial dilutions of soil were prepared. Bacteria were isolated by spreading 1 mL of each

V. P. Ranjan · S. Goel (✉)
Civil Engineering Department, Indian Institute of Technology Kharagpur, Kharagpur, West Bengal, India

D. Sengupta et al. (eds.), *Treatment and Disposal of Solid and Hazardous Wastes*,
https://doi.org/10.1007/978-3-030-29643-8_1

dilution on sterile nutrient agar plates. Soil isolates obtained were tested for its ability to degrade the flower waste. The flower waste was collected from the different temples, it was dried and 1% sterile, dried powder of this flower waste was added in the minimal medium (without carbon source. Therefore, proper and ecofriendly processing of floral waste in these cities and towns is necessary.

The most popular options for waste disposal are open dumping, landfilling, combustion, and composting. Open dumping of MSW results in the degradation of organic materials and creates an environmental nuisance. For a waste with high organic and high moisture content like the floral waste, composting or biogas generation are the only possible treatment options (Goel 2017). Composting is a good treatment option because it is a simple and natural process where organic wastes like remains of plants, garden, and kitchen waste are turned into compost which is a good soil conditioner and fertilizer supplement. It is also the most popular process in India with more than 100 composting companies currently operating in the country (Hiremath and Goel 2017). Reactor-based composting requires knowledge of the rates and extent to which different organic fractions of MSW can be biodegraded.

Different organic fractions of waste like kitchen waste, floral waste, and paper have different extents and rates of biodegradability, which are affected by all factors that influence the growth of microbes. Microbial growth is affected by moisture content, pH, temperature, nutrient availability and chemical composition of the waste fraction. The main objective of this study was to measure the extent and rate of biodegradability of floral waste under optimum aerobic conditions and to relate them to their physico-chemical parameters. Results of floral waste characterization and its biodegradability are presented in this paper.

2 Materials and Methods

2.1 Sample Collection

Four different types of flower waste were chosen for this study: Marigold, Hibiscus, Plumeria (*Champaor Frangipani*) and Ixora (*Rangan*). These flowers were chosen since they are used most frequently in worship, ceremonies, and festivals. Hibiscus and Marigold were collected from Technology Market, IIT Kharagpur. Plumeria and Ixora were collected from the ground in front of Civil Engineering Building, IIT Kharagpur. All samples were dried and shredded in a mixer-grinder to the powdered form. Shredding was done to homogenize the sample for further analysis. For batch biodegradation studies, equal proportions by weight of the four types of flowers were mixed.

Table 1.1 Parameters and analytical methods used for characterizingfloral waste samples

Parameter	Method	Reference/Instrument
Moisture content	Oven method	Worrell and Vesilind (2012)
Calorific value	Bomb calorimeter	Worrell and Vesilind (2012)
Total solids (TS), Volatile solids (VS) and Fixed solids (FS)	Method 2540G; gravimetric	APHA et al. (2005)
Tannin-lignin	Method 5550B; spectrophotometric	APHA et al. (2005)
Total organic carbon (TOC)	Combustion method	O.I. Analytical (2002)
Chemical oxygen demand (COD)	Method 5220 D; spectrophotometric	APHA et al. (2005)
Phosphate	Method 4500-P:D; spectrophotometric	APHA et al. (2005)
Elemental (CHNS-O) analysis	Combustion method	Euro vector EA3000 elemental analyzer

2.2 *Sample Characterization*

Each sample was analyzed in the laboratory for the parameters listed in Table 1.1. Standard methods were used where possible, and any modifications made to the methods are noted in the section on analytical methods (APHA et al. 2005).

2.3 *Batch Biodegradation Studies*

Two long-term batch biodegradation studies were performed under aerobic conditions by changing microbial seed (soil inoculum and cow dung supernatant) and aeration conditions to determine the extent and rate of biodegradation of flower waste. This procedure is similar to the batch biodegradation studies conducted with different organic fractions of municipal solid waste (Kandakatla et al. 2013). A description of the procedure followed for these experiments is provided in Fig. 1.1 and Sect. 2.3.3.

2.3.1 Mineral Media

Mineral media was added to each bottle to ensure that carbon from the solid waste sample was the only limiting nutrient and all other nutrients were more than microbial requirements. The mineral media also had high phosphate concentrations

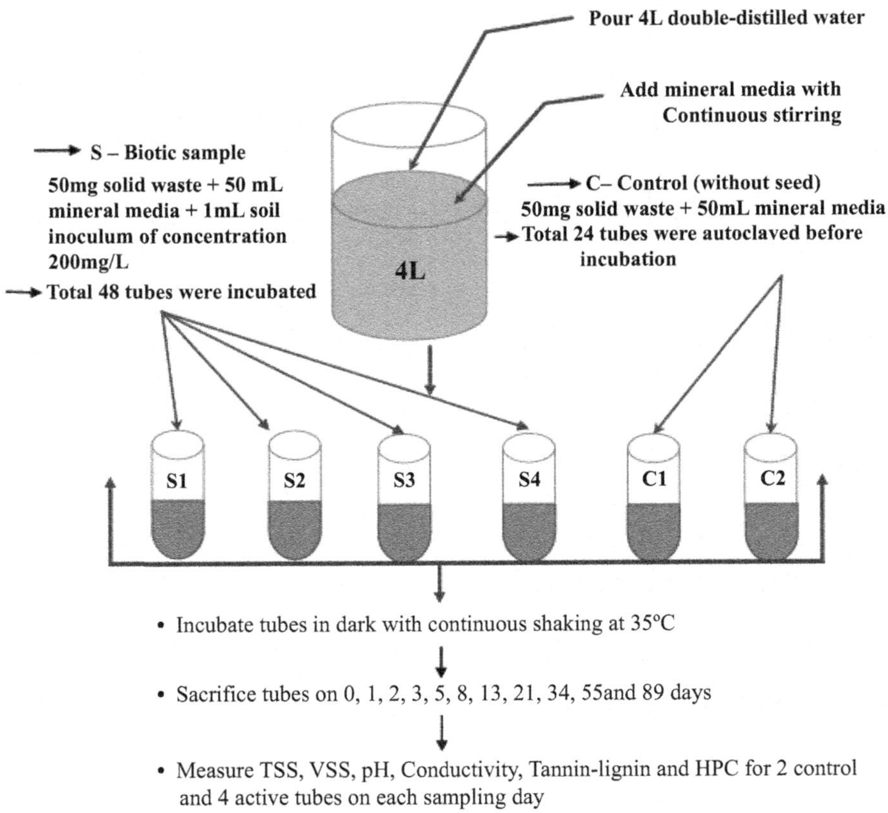

Fig. 1.1 Procedure for setting up first batch biodegradation experiments

to provide a buffer at pH 7.2 throughout the incubation period. Nutrient concentrations to be added were determined based on the standard BOD dilution water composition, the amount of organic material added to each test tube, Method 5210B (APHA et al. 2005) and to ensure that there is no precipitation in the mineral media/dilution water.

Four litres of autoclaved double-distilled water was taken in a large beaker and the following nutrients were added: KH_2PO_4–0.32 g/4 L, K_2HPO_4–0.88 g/4 L, $Na_2HPO_4.7H_2O$–0.7 g/4 L, $MgSO_4.7H_2O$–1.6 g/4 L, NH_4Cl–0.72 g/4 L, $CaCl_2$–0.12 g/4 L, $FeCl_3.6H_2O$–0.016 g/4 L, and Peptone–0.96 g/4 L (for providing trace nutrients).

2.3.2 Microbial Inocula

For microbial seed preparation, 1 g of soil or cow dung was mixed with 10 mL of phosphate-sterilized buffer and was made to 100 mL. This sample was vigorously stirred in a 200 mL beaker with the help of a magnetic stirrer for 1 h for proper mixing of sample and then allowed to settle for half an hour. One mL of supernatant solution was then poured into each of the biotic aerobic tubes to provide microbial seed for biodegradation.

2.3.3 Procedure for Batch Biodegradation Experiments

Two long-term batch biodegradation studies were conducted under aerobic conditions. In the first study, test tubes with vented caps were used. These caps allow air to pass in and out of the test tubes to maintain aerobic conditions during incubation. The vents (or the tortuous air path) prevent contamination by bacteria present in the air outside the test tubes. Soil bacteria were used as inocula in the first study. In the second study, small BOD bottles of volume 125 mL made of borosilicate glass with glass stoppers were used. A water seal was maintained throughout the incubation period to prevent reaeration of the samples during incubation. Cowdung was used as inocula in the second study. All glassware and other materials were autoclaved prior to use.

The procedure for setting up both studies is described here and is shown schematically in Fig. 1.1.

1. A total of 72 autoclaved test tubes of 70 mL capacity were used in batch biodegradation studies of floral waste. Quadruplicate sets of aerobic bioactive tubes and duplicate sets of controls were prepared for study.
2. 50 mg of mixed flower samples were added to each of the aerobic bioactive tubes along with the mineral media and microbial seed and capped. For the control set, 50 mg of sample and 50 mL of mineral media without seed were added to 24 tubes and then autoclaved to ensure that there were no bacteria in these tubes.
3. All tubes were mounted inside an incubator-shaker at a temperature of 35 °C with continuous rotation at 110 rpm. The samples were incubated in the dark to prevent photosynthetic activity and the test tubes are shown in Fig. 1.2.
4. Quadruplicate aerobic biotic tubes were sacrificed along with two control tubes and analyzed on the following days: 0, 1, 2, 3, 5, 8, 13, 21, 34, 55, and 89 days. Each of the sacrificed tubes was analyzed for following parameters: TSS, VSS, pH, DO, conductivity, tannin-lignin, and HPC.

2.4 *Analytical Methods*

1. Moisture content

25 g of each floral sample was taken in a separate crucible dish, and dried to a constant weight in an oven at 105 °C, cooled in a desiccator up to room temperature and the weight of the dried sample was measured. Percentage weight loss of sample in drying gives the moisture content of that sample (Worrell and Vesilind 2012).

Fig. 1.2 Incubated test tubes

2. Calorific value

A bomb calorimeter was used to determine the calorific value of floral samples which provides the gross calorific value at constant volume (Worrell and Vesilind 2012).

3. Tannin-lignin.

Method 5550B, which is a spectrophotometric method was used for Tannin-Lignin determination (APHA et al. 2005). Oven-dried flower samples (1 g) were digested with 50 mL of HCl acid for 12 h, the samples were filtered through Whatman Grade 1 filter paper (pore size 11 μ) and made to 100 mL with double-distilled water. 1 mL of Folin phenol reagent and 10 mL of carbonate tartrate buffer were added in 50 mL of samples for colour to develop. The absorbance of the sample was measured at 700 nm.

4. Total solids (TS), Volatile solids (VS) and Fixed solids (FS)

TS, VS and FS of the flower waste were determined based on Method 2540G (APHA et al. 2005). 25 g of the raw samples were taken for analysis of Total solids, and 1 g of the oven dried samples were taken for analysis of VS and FS.

5. Total Kjeldahl nitrogen (TKN)

Kjeldahl distillation apparatus was used to determine TKN, which is based on Method 4500B (APHA et al. 2005). TKN includes ammonia/ammonium nitrogen and organic nitrogen only. 0.3 g of oven dried sample was mixed with 200 mL of ammonia-free water and poured into a Kjeldahl flask for NH_3/NH_4^+ determination and the residue remaining after determination of NH_3/NH_4^+ was used for organic nitrogen measurement.

6. Chemical oxygen demand (COD)

COD is a measure of the oxygen required to oxidize soluble and particulate organic matter present in water. 1 g of oven-dried sample was taken and digested with 50 mL HCl acid overnight at room temperature. The sample was then filtered through Whatman Grade 1 filter paper (pore size 20 μ) and made to 100 mL with double-distilled water. Filtered sample (2.5 mL) was then subjected to the standard COD test based on Method 5220D (APHA et al. 2005).

7. Phosphate

Method 4500P:D, another spectrophotometric method, was used for phosphate determination (APHA et al. 2005). Oven-dried flower sample (1 g) was digested with 50 mL of HCl acid for 12 h, samples were filtered through Whatman Grade 1 filter paper (pore size 11 μ) and made to 100 mL with double-distilled water. Molybdate (4 mL) and 0.5 mL of $SnCl_2$ reagents were added in 100 mL of sample to develop colour. The absorbance of the samples was measured at 690 nm.

8. Total organic carbon (TOC)

TOC was measured using the Aurora 1030 W TOC Analyzer (O.I. Analytical model 1020A USA) which uses the USEPA-approved combustion method for analyzing samples containing 2 ppb to 30,000 ppm of organic carbon. Oven-dried flower samples (1 g) were digested for 12 h with 50 mL HCl acid, filtered through Whatman filter paper of grade 41 (pore size of 20 μ) and made to 100 mL with double-distilled water. Aqueous samples were introduced into the TOC Analyzer where it was heated to 680 °C in an oxygen-rich environment. Water is evaporated and removed from the system, and the total carbon content from the sample is oxidized to form carbon dioxide. The carbon dioxide generated by oxidation is detected using Non-Dispersive Infra-Red gas analyzer (NDIR) which is directly proportional to the amount of organic carbon in the sample (O.I. Analytical 2002).

9. pH

pH was measured using a digital pH meter by placing the pH electrode in the sample solutions. The instrument was calibrated with known standard buffer solutions of pH 4 and 7 before measuring pH of the samples.

10. Conductivity

Conductivity is a measure of the ability of an aqueous solution to conduct an electric current and is proportionate to the ion concentration in solution. A digital conductivity meter was used for conductivity measurements by placing an electrode in the sample solutions. The instrument has to be calibrated with known standards of conductivity of 1000 μS/cm or 2000 μS/cm before measuring conductivity of sacrificed samples.

11. Dissolved oxygen (DO)

Dissolved oxygen in the samples were determined by a DO meter (Lutron, Model: PDO-519). Incubated test tubes were first vortexed and a small volume poured in 50 mL beakers. The DO meter was first calibrated with air and the tip of the probe was immersed in the sample and shaken gently to get a stable reading.

12. Total suspended solids (TSS)

TSS of the samples were analyzed based on Method 2540D (APHA et al. 2005). TSS was determined by filtering both sets of samples, biotic and control sets through 934-AH glass microfiber filter having diameter 47 mm and pore size 1.5 μm. TSS is defined as residue left upon evaporation at 103–105 °C in a hot air oven until constant weight is obtained.

13. Volatile suspended solids (VSS)

VSS of the samples were determined based on Method 2540E (APHA et al. 2005). It is defined as the weight of TSS lost during ignition at 550 °C. VSS is a direct indicator of the combustible organic matter contained in the samples.

14. Heterotrophic plate counts (HPC)

Heterotrophic plate counts were determined on some sampling days for each test tube during biodegradation. A known volume of diluted sample was poured and spread with the help of a spreader on nutrient agar-filled Petri plates/dishes. The Petri dishes were incubated upside down in an incubator at 35 °C for 24 h before counting of plates. In a plate count, the numbers of colony forming units (cfu) were determined, and it was assumed that each colony represents at least one cell. The initial sample was serially diluted to obtain a small but statistically significant number of colonies on each plate. A statistically significant number of colonies on a standard Petri plate was taken as 30–300 cfu for counting (APHA et al. 2005). Dilutions that result in cfu values outside this range were not considered acceptable.

15. Elemental (C, H, N, S and O) analysis

Euro Vector EA3000 Series of CHNS-O Elemental analyzer was used to determine elemental analysis of floral samples, which gives elemental composition in terms of percentage by weight. Floral samples were weighed in mg in a tin capsule and then dropped in to the combustion chamber for complete combustion and CHNS determination. Combustion mixture flows in to the GC column for separation and is detected by a thermal conductivity detector (TCD) in the sequence: N_2, CO_2, H_2O and SO_2. The results are obtained as %N, %C, %H and %S.

3 Results and Discussion

3.1 Sample Characterization

Four different types of flowers were collected and analyzed for various physical and chemical parameters. A summary of the results is provided in Table 1.2.

(a) **Moisture content:** The natural moisture content of different flowers was very high and varied from 81.72% to 91.2%. Hibiscus had the highest moisture content of 91.2% followed by Marigold (87.12%), Plumeria (85.63%) and Ixora (81.72%). The mixed flower waste had a moisture content of 86.5% and these results are similar to those of rose processing waste which had moisture content of 90.5% (Tosun et al. 2008).

(b) **Total solids:** Ixora (*Rangan*) had the highest total solid contents of 18.28% followed by Plumeria (14.37%), Marigold (12.88%) and Hibiscus (8.8%). This corresponds to the fact that the moisture content of Hibiscus was the highest. The mixed flower waste had a total solids content of 13.52%.

(c) **Fixed solids (FS):** Marigold had the highest fixed solid content of 8.8% followed by Hibiscus (8.08%), Plumeria (6.22%), and Ixora (3.8%). Low fixed solids in flower samples are due to the low inert (inorganic) content.

Table 1.2 Characterization of different types of flowers

Parameters	Marigold	Hibiscus	Plumeria (Champa)	Ixora (Rangan)	Mixed flower
Moisture content (%)	87.12	91.2	85.63	81.72	86.48
Total solids (%)	12.88	8.8	14.37	18.28	13.52
Fixed solids (% of TS)	8.80	8.08	6.22	3.8	6.78
Volatile solids (% of TS)	91.20	91.92	93.78	96.2	93.22
Tannin-lignin (mg/g of waste)	65.5	26.7	24.2	7.1	29.8
Calorific value (kJ/kg)	17,808	16,778	16,886	17,601	17,512
Total carbon (% C by wt.)	50.008	49.398	45.354	51.689	52.837
Total nitrogen (% N by wt.)	3.602	4.891	2.702	4.233	4.468
C/N ratio	13.89	10.10	16.785	12.211	11.825
COD (mg O_2 /mg of waste)	1.259	1.011	1.327	1.823	1.366
Total organic carbon (% C by wt.)	27.36	28.03	23.09	24.196	24.69
Phosphate (mg/g)	1.66	0.533	0.455	2.17	1.20
TKN (mg/g waste)	10.742	9	2.9	3.4	6.55

(d) **Volatile solids (VS):** Ixora had the highest volatile solid contents of 96.2% followed by Plumeria (93.78%), Hibiscus (91.92%) and Marigold (91.2%). High volatile solids are due to the presence of high organic content in the samples.
(e) **Calorific value:** Plumeria had the highest calorific value of 16,886 kJ/kg followed by Marigold (17,808 kJ/kg), Ixora (17,601 kJ/kg) and Hibiscus (16,778 kJ/kg). Calorific values represent the amount of heat produced during combustion. This is another measure of the organic content of a bulk sample.
(f) **Total Kjeldahl nitrogen (TKN):** Marigold had the highest TKN value of 10.742 mg/g of waste, followed by Hibiscus (9 mg/g), Ixora (3.4 mg/g) and Plumeria (2.9 mg/g). TKN values are a measure of the organic nitrogen and ammonia content of the samples.
(g) **Phosphate content:** Ixora had the highest phosphate content of 2.17 mg/g of waste, followed by Marigold (1.66 mg/g), Hibiscus (0.533 mg/g) and Plumeria (0.455 mg/g).
(h) **Tannin-lignin content:** Marigold had the highest tannin-lignin content of 65.5 mg/g of waste followed by Hibiscus (26.7 mg/g), Plumeria (24.2 mg/g), and Ixora (7.1 mg/g). Tannin and lignin are naturally occurring organic components of plant cell walls representing 5–10% of the dry wt. of leaves and needles (Klotzbücher et al. 2011). Lignin is aromatic and is often considered to be more resistant to biodegradation compared to aliphatic compounds.
(i) **COD content:** Ixora had the highest COD values of 1.823 mg of O_2/g of waste followed by Plumeria (1.327 mg/g), Marigold (1.259 mg/g), and Hibiscus (1.011 mg/g). Higher the value of COD, higher will be the requirement for oxygen during decomposition of organic matter.

3.2 First Batch Biodegradation Study

Of the two long-term studies conducted, the first batch biodegradation study with floral waste is described in detail here, and the results of both studies are summarized in Table 1.7. Biodegradation of floral waste was evaluated over 89 days of incubation by sacrificing two control and four active sample tubes on each sampling day. Various parameters were analyzed for each sampling day, and the results of each parameter are discussed here:

1. pH

It was observed that initial pH varied from 6.918 ± 0.103 to 7.103 ± 0.118 at the end of 89 days of incubation. Minimum pH was 6.56 in the controls and 6.6 in the biotic tubes after 13 days of incubation. A slight decrease in pH for both biotic and control sets was noted at the beginning of biodegradation due to the production of organic acids. These results demonstrate the effectiveness of the phosphate buffer in maintaining near neutral pH throughout the incubation period (Fig. 1.3).

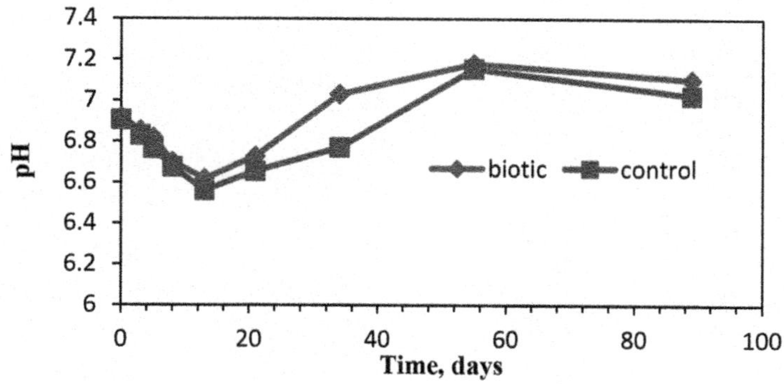

Fig. 1.3 Variation in pH during incubation

2. Conductivity

Conductivity of both biotic and control sets of solutions increased constantly over time. For the biotic set, initial conductivity increased from 1367.5 ± 31.5 μS/cm to 1516.25 ± 6.75 μS/cm on the 89th day. In the control set, the initial conductivity was 1329.5 ± 12.5 μS/cm and increased to 1499 ± 10 μS/cm after 89 days. Increase in conductivity indicates an increase in ionic concentration that can be attributed to solubilization of solid samples. Variation in conductivity is shown in Fig. 1.4.

3. Tannin-lignin

Initially, the tannin-lignin content of the biotic set of samples was observed to be 32.73 ± 1.48 mg/L, and after 89 days, it was 9.65 ± 1.53 mg/L. In the control set, the initial concentration was 31.20 ± 0.67 mg/L, and after 89 days, it was 25.28 ± 0.29 mg/L. Variation in tannin-lignin is shown in Fig. 1.5. These results indicate the degradation of tannin-lignin in the biotic and control samples and demonstrate the greater rate and extent of biodegradation in the biotic tubes versus the controls. The reason for the decrease in concentration of tannin-lignin in the biotic set is that microbes breakdown large molecular weight compounds like lignin initially and produce smaller-sized aromatics that can be imported into the cell for further catabolism (Brown and Chang 2014).

4. Total suspended solids (TSS) and Volatile suspended solids (VSS)

TSS present in solution decreased due to degradation of floral waste. Initial TSS concentration for the biotic set was 0.57 ± 0.02 g/L and decreased to 0.37 ± 0.02 g/L after 89 days of degradation (Fig. 1.6). This represents 35.08% degradation in TSS concentration. For the control set, initial concentration was 0.44 ± 0.05 g/L and decreased to 0.38 ± 0.01 g/L after 89 days, and this represents 13.63% degradation of the solid sample. The difference in initial TSS concentrations between biotic and control samples indicates mineralization of these samples due to autoclaving. These results also indicate that only 35% of the TSS was biodegradable and complete biodegradation was achieved in 55 days.

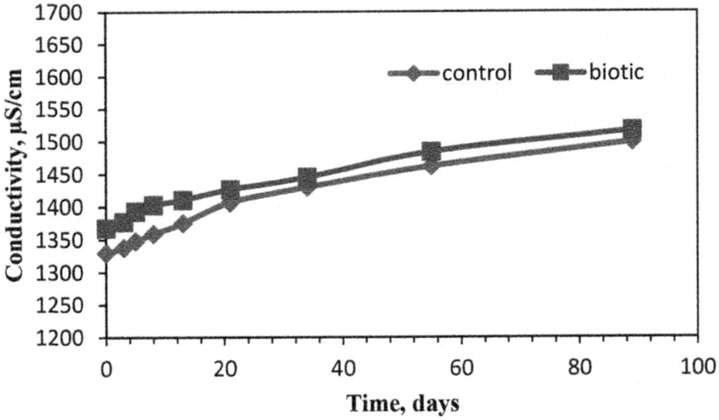

Fig. 1.4 Variation in conductivity during incubation

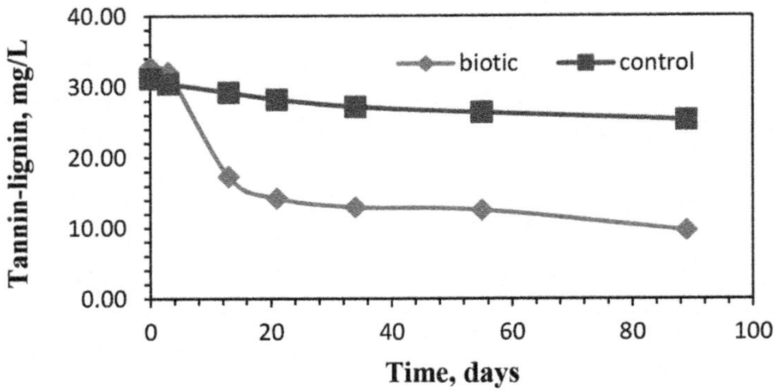

Fig. 1.5 Change in tannin-lignin concentration during incubation

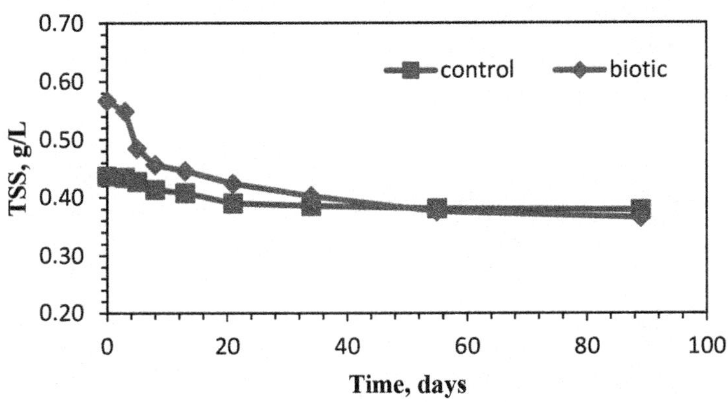

Fig. 1.6 Variation in TSS during incubation

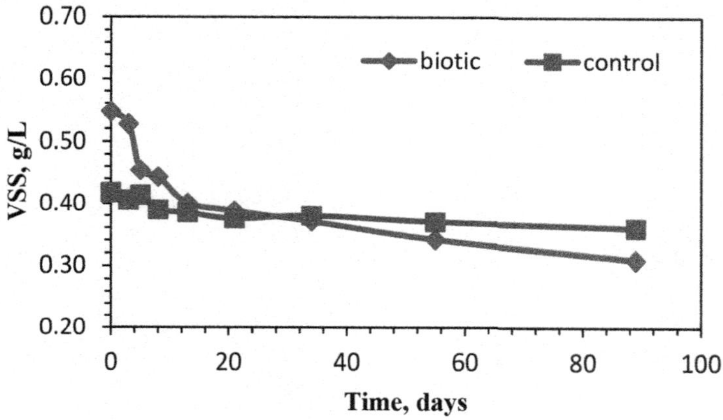

Fig. 1.7 Variation in VSS during incubation

Initial VSS concentration for the biotic set was 0.55 ± 0.02 g/L and decreased to 0.31 ± 0.03 g/L after 89 days of degradation (Fig. 1.7). This represents 43.64% degradation in VSS concentration. For the control set, initial concentration was 0.46 ± 0.02 g/L, and final concentration was 0.39 g/L after 89 days of incubation and represents 15.21% decrement. Significantly lower reduction in TSS and VSS in control sets compared to biotic samples indicates that reduction in the autoclaved control set was mainly due to abiotic mineralization.

5. Dissolved oxygen (DO)

Initial DO available in the biotic sample was 6.15 ± 0.15 mg/L, and after 89 days, it decreased to 5.71 ± 0.08 mg/L. This represents 7.15% decrement in DO concentration. Decrement in DO in the biotic set of the sample was due to the presence of microbes which degraded floral waste by consuming DO. For the control set, the initial concentration of DO was 6.35 ± 0.05 mg/L, and after 89 days, final DO was 6.12 ± 0.06 mg/L, which shows 3.62% reduction in DO concentration (Fig. 1.8). In the control set, there was little or no change in DO due to the absence of microbial and both sets were subjected to continuous shaking under aerobic conditions. Further, the open vents in the test tube caps allowed oxygen to enter these test tubes during incubation contributing to relatively high DO levels.

6. Heterotrophic plate count (HPC)

The microbial concentration (cfu/mL) of floral waste in the biotic samples was measured on the 1stday and found to be 4.63×10^6 and after 89 days, it was 4.63×10^8 indicating an increase in concentration by two logs or 100-fold. In the control set, very few bacterial colonies were found; the average HPC was 100 cfu/mL and is attributed to low levels of contamination. Variations in bacterial concentrations during incubation are shown in Fig. 1.9.

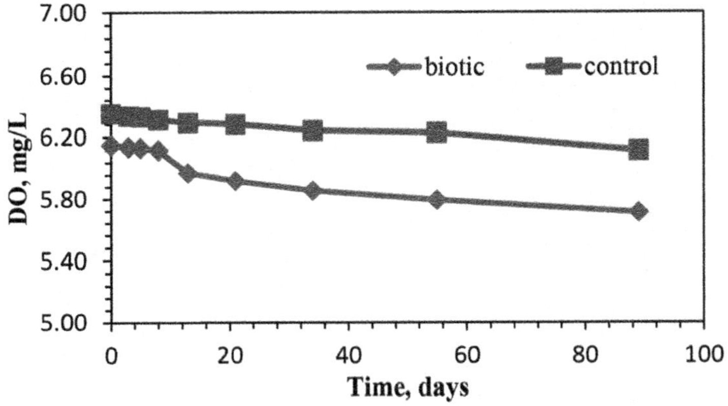

Fig. 1.8 Variation in DO during incubation

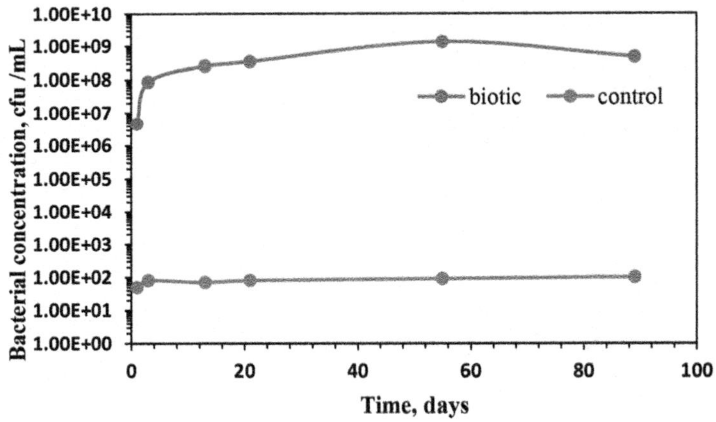

Fig. 1.9 Variations of bacterial concentration during incubation

7. Visualization of flowers by SEM analysis

All flowers were dried in a hot air oven, and ground to powdered form and then taken for Scanning Electron Microscopy (SEM). SEM was used for viewing the surface features of the flower waste under magnification ranging from 500× to 1500×. SEM allows morphological characterization of the surface features of floral structures in a manner that retains the topography or three-dimensional appearance of the structure. The original surface topography of different flowers is shown in Fig. 1.10. In SEM of Hibiscus, flat materials and ball-like structures with spines are visible. Based on comparison with other images, the ball-like structures are likely to be pollen grains, and the flat materials are petals. After 39 days of incubation, the appearance of samples from both biotic and control sets are shown in Fig. 1.11. No

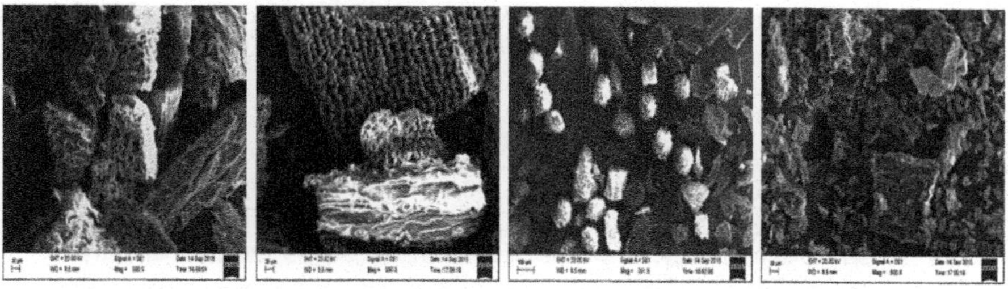

a) Marigold b) Ixora c) Hibiscus d) Plumeria

Fig. 1.10 Appearance of different flowers

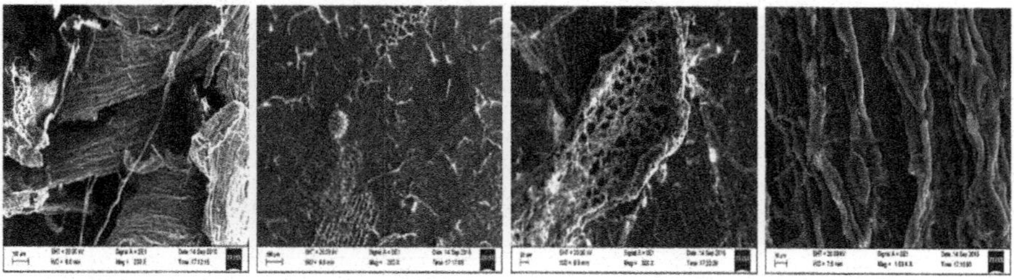

a) Control (200X) b) Biotic (200X) c) Biotic (500X) d) Control (1.50K X)

Fig. 1.11 Appearance of flower sample in control set and biotic setafter 39 days of incubation

significant change in the appearance of the biotic sample was observed after 39 days of incubation. Pollen grains were not degraded even after 39 days of incubation while some parts of petals indicate degradation.

8. FTIR analysis of floral sample

FTIR spectroscopy can be used to confirm the decomposition of polypeptides, polysaccharides, aliphatic, aromatic, carboxylic, phenolic groups, and lignin during the biodegradation process (Bhat et al. 2017). It is one of the most reliable techniques for determining change in specific functional groups. FTIR spectra of initial and final biotic samples were obtained in the mid-infrared range of 4000–400 cm^{-1} and the percentage transmission through the biotic sample was greater than the control sample indicating the impact of biodegradation on the sample. Specific changes were observed in bands corresponding to different types of chemical bonds as shown in Table 1.3 and Fig. 1.12.

FTIR analysis of the degraded sample follows the same trend as of the control sample, but significant structural changes can be observed. Distinct differences between the biotic and control samples were observed at 3384.46, 2917.77, 1646.91, 1442.49 and 1058.73 cm^{-1} (Fig. 1.12). The greatest differences in transmittance were observed in the fingerprint region between 600 and 1400 cm^{-1} indicating biodegradation of polysaccharides to the greatest extent (1000–1100 cm^{-1}) and aromatic compounds like lignin (1430–1470 cm^{-1}) to

Table 1.3 Characteristic bands of the FTIR spectra of floral waste (based on Liu et al. 2014; Bhat et al. 2017)

Wave number (Range)	Functional group	Control (Wave number, % Transmittance)	Biotic (Wave number, % Transmittance)	Change in % transmittance
3300–3500	O-H stretching of phenols and carboxylic group	3384.459, 67.5726	3413.386, 74.59505	7.03
2850–2925	C-H band in aliphatic methylene groups of lipids and fats	2917.77, 73.48534	2931.269, 79.32641	5.82
1640–1655	C=O stretching of carboxylates, C=N stretching of amide, and C=C stretching of alkene and aromatic ring mode	1646.911, 67.70178	1643.054, 79.06642	11.36
1430–1470	Asymmetric bending in CH_3 (lignin)	1442.493, 73.99207	1438.636, 83.57227	9.58
1000–1100	C-O stretching of polysaccharides	1058.728, 54.70376	1079.941, 80.02473	25.32

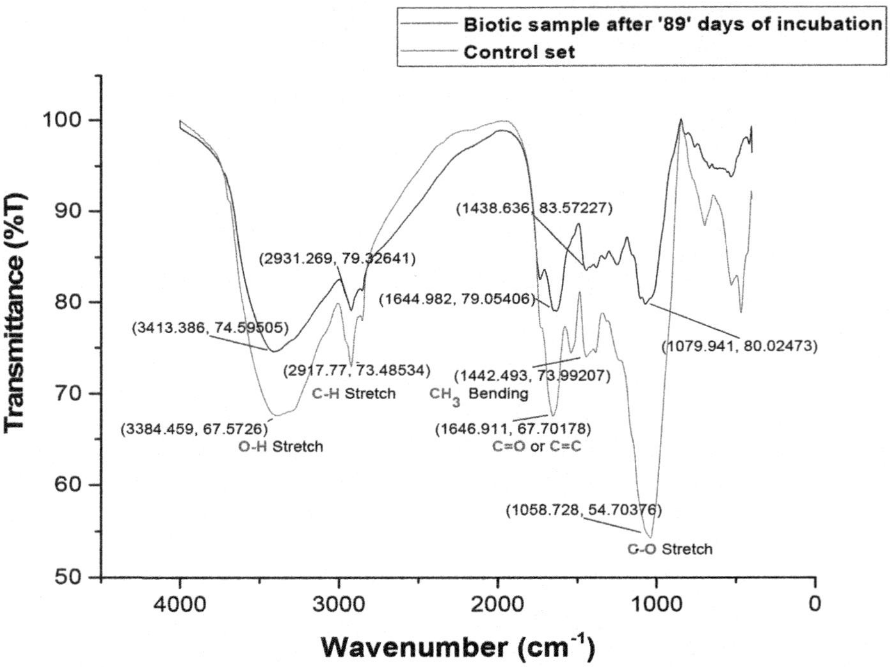

Fig. 1.12 FTIR analysis of control set and biotic set after 89 days of incubation

some extent. The C-H stretching vibration which corresponds to 2917.77 cm^{-1} from aliphatic methylene groups of lipids and fats showed the smallest change indicating least biodegradation of this group of compounds. The band corresponding to 1646.91 cm^{-1} is due to COO^- and C=O vibrations of amides, ketones, and quinines (Jouraiphy et al. 2005; Hajjouji et al. 2007). The broad and strong vibration located at wavenumber 3384.46 cm^{-1} indicates the presence of O-H bond stretching due to phenols and carboxylic groups which may be associated with lignin. The change in % transmittance values of almost all functional groups as shown in Fig. 1.12 proves that some degradation has happened in the biotic sample as microorganisms play an important role in the degradation of various organic wastes. Microbes are one of the principal sources of enzymes like cellulase and amylase which are responsible for breaking down large, complex macromolecules like lignin and cellulose in the biotic samples as compared to the control set (Awasthi et al. 2018).

9. Carbon/nitrogen (C/N) ratio

The C/N ratio was reduced from 11.82 to 7.33, and from 11.82 to 10.49 for biotic and control sets, respectively. Significant reduction in C/N ratio in the biotic set indicates loss of organic carbon due to biological oxidation of organic compound and release of CO_2 leading to an increase in the proportion of total nitrogen in the medium (Loubna et al. 2014).

3.3 Determination of Reaction Rate Constants

Reaction rate constants were determined using the results obtained from batch experiments by plotting graphs for zero-order reaction, first-order reaction, second-order reaction and saturation reaction kinetics. Regression coefficients (R^2) obtained from the graphs plotted for various reaction orders were used to determine the best-fit for the batch data. TSS and VSS concentrations were used to determine the reaction rate constants.

Initial TSS concentrations in batch biodegradation studies were 0.57 g/L. The graphs for different reaction orders were plotted to check the best-fit models and are shown in Figs. 1.13 and 1.14. R^2 values for zero-order, first-order, second-order and saturation reaction (mixed order kinetics) were found to be 0.6506, 0.7038, 0.7547 and 0.9997 respectively and are shown in Table 1.4. Maximum R^2 value and, therefore, best-fit are reported for saturation kinetics.

Initial VSS concentration in batch biodegradation studies was 0.55 g/L. R^2 value for zero-order, first-order, second-order and mixed-order was found to be 0.7041, 0.7791, 0.848 and 0.9993, respectively, implying that the best-fit was for saturation kinetics again. The graph plotted for different orders of kinetics versus time and graph for mixed-order kinetics are shown in Figs. 1.15 and 1.16 respectively. R^2 value and rate constant (k) for different reaction orders are shown in Table 1.5.

3.4 Second Batch Biodegradation Study

The second batch study was conducted for the same mixed floral waste sample by changing microbial seed from soil inoculum to cow dung. Small-sized BOD bottles

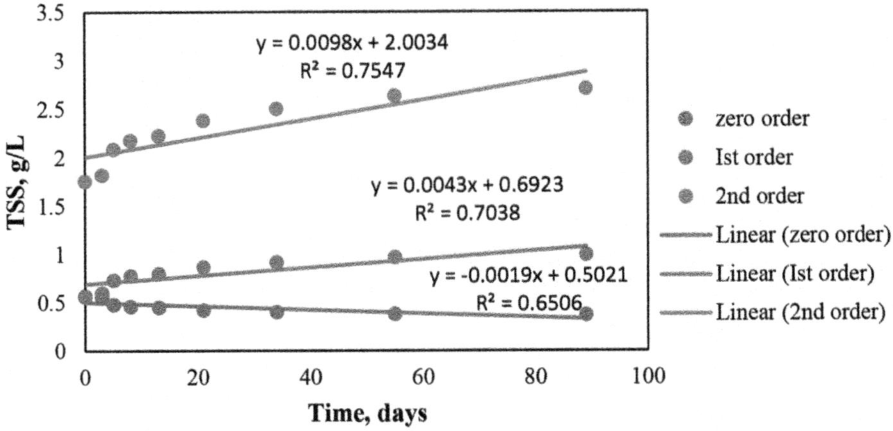

Fig. 1.13 Kinetic rate constants for degradation of TSS

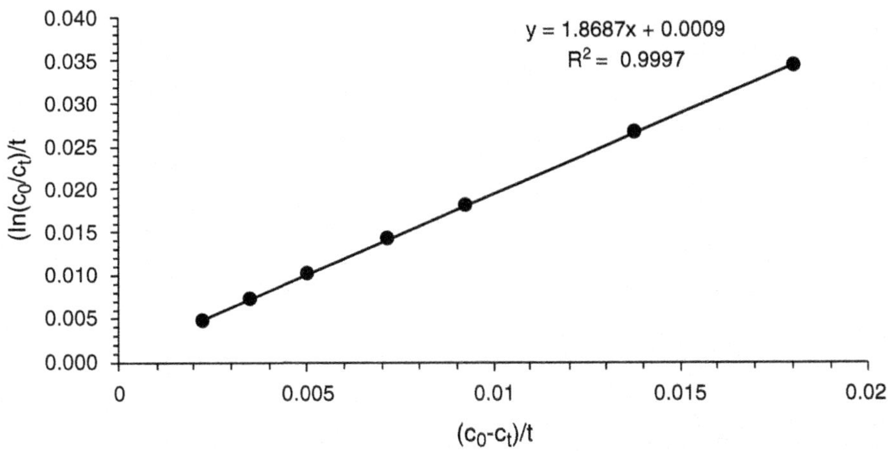

Fig. 1.14 Saturation type reaction based on TSS

Table 1.4 R^2 for velocity constant (K) and 'k' values for different reaction orders and TSS

	Zero order ($g\ L^{-1}\ day^{-1}$)	1st order (day^{-1})	2nd order ($Lg^{-1}\ day^{-1}$)	Saturation kinetics
R square	0.6506	0.7038	0.7547	0.9997
k	0.0019	0.0043	0.0098	0.000482
K (g/L)				0.53513

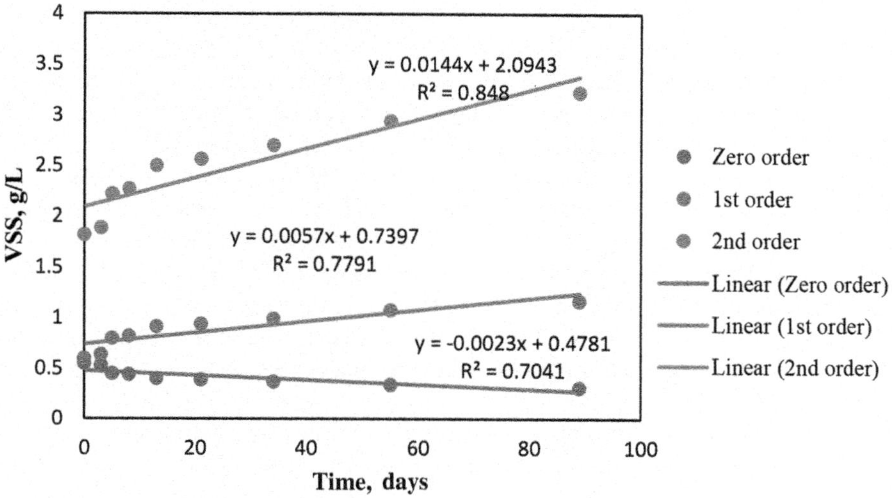

Fig. 1.15 Kinetic rate constants for degradation of VSS

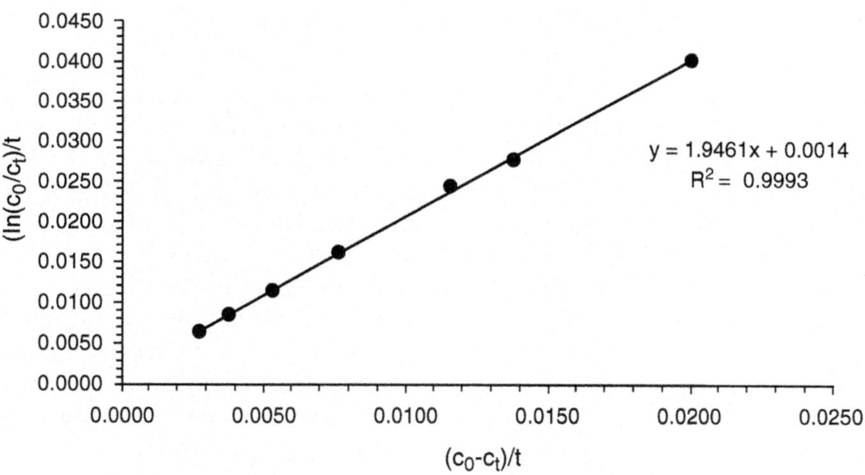

Fig. 1.16 Saturation type reaction based on VSS

Table 1.5 R^2, velocity constant (K) and 'k' values for different reaction orders and VSS

	Zero order ($g\ L^{-1}\ d^{-1}$)	1st order (day^{-1})	2nd order ($Lg^{-1}\ d^{-1}$)	Saturation kinetics
R square	0.7041	0.7791	0.848	0.9993
k	0.0023	0.0057	0.0144	0.000719
K, g/L				0.51384

were used in this study instead of test tubes with vented caps so that no reaeration could occur during incubation. TSS removal was observed to be 41.38%, and VSS removal was observed to be 44.64% after 89 days of incubation. No significant change in pH was observed during 89 days of incubation in the biotic set as well as in control set. Initially, DO concentration for the biotic set was observed to be 6.53 mg/L, and after 89 days of incubation DO concentration was reduced to 5.16 mg/L, which shows 7.15% reduction in DO concentration. DO concentration in the control set remained almost constant and showed only 3.62% decrement after 89 days of incubation. The significant decrement in tannin-lignin concentration was observed in the biotic set after 89 days of incubation. Tannin-lignin removal was observed to be 72.05% in biotic sets after 89 days. The bacterial concentration (cfu/mL) of floral waste in the biotic samples was measured on the first day and found to be 2.17×10^7, and after 89 days, it was 3.13×10^6.

The kinetics of biodegradation were analyzed for the two studies, and maximum R^2 values were obtained for saturation or mixed-order kinetics as shown in Table 1.6.

3.5 *Stoichiometry of Aerobic Biodegradation*

Elemental composition (C, H, N and O) of four different types of floral samples were determined on a percentage by weight basis. Sulphur was not taken in to consideration in the elemental composition because its content was negligibly small. Chemical formulas were established (Table 1.7) based on the elemental composition determined for four different types of flowers, and mixed flower waste.

If complete oxidation of each flower sample and the mixed flower waste sample is considered, then it is assumed that the end-products are carbon dioxide, water, ammonia, sulphate and phosphate. No new biomass or residual organic matter is present in such cases. Stoichiometric equations for complete oxidation of floral waste are shown in Table 1.8. The assumption in these equations is that the substrate is completely oxidized and the amount of oxygen consumed is dependent only on the initial elemental composition of the waste. Oxygen required (in moles) for complete mineralization of each flower sample and mixed flower samples can be noted from these equations. They represent the theoretical oxygen demand for each sample. Based on this stoichiometry, 0.95 g O_2/g waste is required for complete oxidation of the mixed floral waste.

Under the experimental conditions of this study, the mixed flower samples were only partially oxidized after 89 days of incubation and the residual material which included organic waste and new biomass were subjected to elemental analysis. Another stoichiometric equation was formulated based on the elemental analyses of the waste material before and after incubation and is shown below. The empirical formula for the mixed flower waste sample is on the left hand side and the residual material after incubation is on the right hand side of the reaction.

Table 1.6 Comparison of biodegradation studies between floral waste and leaf litter

Parameter		Floral waste				Leaf litter (Kandakatla et al. 2013)	
Microbial seed added		Soil inoculum[a]		Cow dung supernatant[a]		Soil inoculum[b]	
		Biotic (n = 4)	Control (n = 2)	Biotic (n = 4)	Control (n = 2)	Biotic (n = 3)	Control (n = 1)
pH	Initial	6.92 ± 0.11	6.905 ± 0.015	6.58 ± 0.05	6.57	6.94 ± 0.06	6.95
	Final	7.11 ± 0.12	7.025 ± 0.095	6.31 ± 0.19	6.40	6.98 ± 0.13	7.02
Conductivity	Initial	1367.5 ± 31.5	1329.5 ± 12.5	1377.67 ± 8.33	1386	1460 ± 300	1300
	Final	1516.25 ± 6.75	1499 ± 10	1991.67 ± 18.33	1595	4540 ± 150	3900
	Increment	**9.81%**	**11.30%**	**44.57%**	**15.08%**	**210.95%**	**200%**
DO (mg/L)	Initial	6.15 ± 0.15	6.35 ± 0.05	6.53 ± 0.58	7.2	NA	NA
	Final	5.71 ± 0.08	6.12 ± 0.06	5.16 ± 0.31	6.6	NA	NA
	Decrement	**7.15%**	**3.62%**	**20.98%**	**8.33%**	NA	NA
TSS (g/L)	Initial	0.57 ± 0.02	0.44 ± 0.05	0.58 ± 0.02	0.55	0.82 ± 0.05	0.71
	Final	0.37 ± 0.02	0.38 ± 0.01	0.34 ± 0.01	0.50	0.2 ± 0.02	0.61
	Degradation	**35.08%**	**13.63%**	**41.38%**	**9.1%**	**75.6%**	**14.08%**
VSS (g/L)	Initial	0.55 ± 0.02	0.46 ± 0.02	0.56 ± 0.02	0.52	0.75 ± 0.03	0.83 ± 0.03
	Final	0.31 ± 0.03	0.39	0.31 ± 0.02	0.48	0.18 ± 0.02	0.68
	Degradation	**43.64%**	**15.21%**	**44.64%**	**7.7%**	**76%**	**18.07%**
O_2 (mg) consumed/VSS (g) degraded		**1.833**	**3.28**	**5.48**	**15**	**NA**	**NA**
Tannin-lignin (mg/L)	Initial	32.73 ± 1.48	31.20 ± 0.67	28.16 ± 2.0	28.84	111.97 ± 14.72	42.85
	Final	9.65 ± 1.53	25.28 ± 0.29	7.87 ± 0.95	19.06	52.23 ± 9.58	20.1
	Removal	**70.51%**	**18.97%**	**72.05%**	**33.91%**	**53.35%**	**53.09%**
HPC (cfu/mL)	Initial	4.63×10^6	50	2.17×10^7	0	5×10^7	3.2×10^7
	Final	4.63×10^8	100	3.13×10^6	40	1.04×10^8	7×10^7

(continued)

Table 1.6 (continued)

Parameter		Floral waste				Leaf litter (Kandakatla et al. 2013)	
Microbial seed added		Soil inoculum[a]		Cow dung supernatant[a]		Soil inoculum[b]	
		Biotic (n = 4)	Control (n = 2)	Biotic (n = 4)	Control (n = 2)	Biotic (n = 3)	Control (n = 1)
C:N ratio	Initial	11.82	11.82	11.82	11.82	NA	NA
	Final	7.33	10.49	6.65	10.75	NA	NA
	Decrement	**37.98%**	**11.25**	**43.74**	**9.05**	NA	NA
Reaction order	Order	Mixed	NA	Mixed	NA	Second	NA
Based on TSS conc.	R^2	0.9302	NA	0.9301	NA	0.9578	NA
	K	0.01438	NA	0.0159	NA	0.0303	NA
Reaction order	Order	Mixed	NA	Mixed	NA	Second	NA
Based on VSS conc.	R^2	0.9258	NA	0.9493	NA	0.975	NA
	K	0.0179	NA	0.0233	NA	0.0369	NA

[a]Aerobic condition with 89 days of incubation, [b]Anaerobic condition with 89 days of incubation, NA = Not available

Table 1.7 Initial elemental composition of floral waste

Types of flower	Elemental composition as per CHNSO elemental analyser	Elemental composition after normalizing for N
Marigold	$C_{50}H_{15.13}O_{31.25}N_{3.60}$	$C_{13.88}H_{4.20}O_{8.68}N$
Hibiscus	$C_{49.40}H_{20.96}O_{24.75}N_{4.89}$	$C_{10.10}H_{4.27}O_{5.06}N$
Plumeria	$C_{45.35}H_{10.96}O_{40.99}N_{2.70}$	$C_{16.79}H_{4.06}O_{15.18}N$
Rangan	$C_{51.69}H_{20.78}O_{23.29}N_{4.23}$	$C_{12.22}H_{4.91}O_{5.50}N$
Mixed flower	$C_{52.84}H_{20.58}O_{31.25}N_{3.60}$	$C_{14.67}H_{5.71}O_{8.68}N$

Table 1.8 Stoichiometric equation for complete oxidation of organic matter

Types of flower	Stoichiometric equation
Marigold	$C_{13.88}H_{4.20}O_{8.68}N + 9.84\ O_2 \rightarrow 13.88\ CO_2 + 0.6\ H_2O + NH_3$
Hibiscus	$C_{10.10}H_{4.27}O_{5.06}N + 7.89\ O_2 \rightarrow 10.10\ CO_2 + 0.63\ H_2O + NH_3$
Plumeria	$C_{16.79}H_{4.06}O_{15.18}N + 9.46\ O_2 \rightarrow 16.79\ CO_2 + 0.53\ H_2O + NH_3$
Rangan	$C_{12.22}H_{4.91}O_{5.50}N + 9.95\ O_2 \rightarrow 12.22\ CO_2 + 0.95\ H_2O + NH_3$
Mixed flower	$C_{14.67}H_{5.71}O_{8.68}N + 11\ O_2 \rightarrow 14.67\ CO_2 + 1.35\ H_2O + NH_3$

$$C_{14.67}H_{5.717}O_{8.68}N + 4.49\ O_2 \rightarrow C_{7.33}H_{4.64}O_{2.44}N + 7.34CO_2 + 0.54\ \ H_2O$$

Based on the above stoichiometric equation, the amount of oxygen required to biodegrade the waste material was 0.43 g O_2/g waste. In the first batch biodegradation study, the test tubes had vented caps implying there was unrestricted availability of oxygen. Therefore, the amount of oxygen consumed could not be estimated. However in the second batch biodegradation study, where the solution was not allowed to reaerate during incubation, the amount of oxygen consumed was estimated to be 5 mg O_2/g VSS of waste biodegraded which is far below the ThOD estimated for this sample. These results suggest that biodegradation in these experiments was done not only by aerobic bacteria but also by facultative and anaerobic bacteria.

3.6 *Summary of Batch Biodegradation Studies*

Major findings in this study with floral waste under aerobic conditions with soil inoculum or cow dung as microbial seed are summarized here and in Table 1.6 and further compared to the results with leaf litter biodegradation (Kandakatla et al. 2013):

1. No significant change in pH was observed in any of these studies. This indicates that the phosphate buffer added was adequate.
2. Conductivity increased in both studies. However, the increase in conductivity, which indicates solubilization of solid material, was much less in the study with

floral waste under aerobic conditions as compared to another study with leaf litter. These results suggest that floral waste solubilizes to a lesser extent than leaf litter.
3. TSS and VSS concentrations decreased significantly in the biotic sets of all three studies. However, the extent of biodegradation of floral waste was much less than that of leaf litter. Also, change in microbial seed and aeration conditions resulted in slightly better TSS and VSS removals with cow dung and no aeration during incubation as compared to soil inoculum and constant aeration in the biotic sets for floral waste. These results are interesting since they show that floral waste under aerobic conditions was less biodegradable than leaf litter. Further, the smaller increase in conductivity during floral waste biodegradation as compared to leaf litter biodegradation confirms this finding.
4. Microbial counts were higher in biotic sets of floral waste after incubation and contamination was controlled to a greater extent in the control set in this case as compared to the leaf litter study.
5. Kinetic rate constants were calculated for TSS and VSS variations observed during both batch biodegradation studies. Coefficients of determination (R^2) values for zero-order and first-order kinetics and second-order kinetics were less than for mixed-order or saturation kinetics. These results indicate that mixed-order kinetics were the best-fit for the degradation of floral waste based on high coefficients of determination (R^2).
6. The initial C:N ratio was found to be 11.82 and after 89 days, the C:N ratio decreased to 7.33 which shows 37.98% reduction in C:N ratio for the biotic sample. This shows mineralization of C from the samples during incubation and is an extremely important outcome for composting where the compost is to be enriched in terms of N content. It also points to the need for mixing C-rich samples like flowers, leaf litter and paper with N-rich wastes like cow dung and wastewater sludge during composting to improve compost quality.

4 Conclusions

(a) No significant differences were observed between the two aerobic batch biodegradation studies with floral waste even after changing microbial seed and aeration conditions.
(b) TSS removals for the first and second batch biodegradation study were 35.08% and 41.38%, and VSS removals were 43.64% and 44.64% after 89 days. In the control set, there was a slight decrement in TSS (13.63% and 9.1%) and VSS (15.21% and 7.7%) due to solubilization of organic sample and mineral media present in the solution.
(c) The conductivity of the samples for both floral waste studies increased constantly during the incubation period of 89 days and can be attributed to solubilization of solid sample material in biotic and control sets.

(d) There was a significant change in C:N ratio from 11.82 to 7.33 or 6.65 in the two studies which is an extremely important outcome for composting where the compost is to be enriched in terms of N content.
(e) FTIR analysis of biotic and control samples showed a decrease in the presence of major functional groups like phenols, carboxylic, methylene, lignin and polysaccharides due to biodegradation of floral waste samples.
(f) The kinetics of batch biodegradation of floral waste under aerobic conditions were found to best-fit a mixed-order kinetic model with high coefficients of determination (R^2). Regression coefficients (R^2) for mixed-order kinetics based on TSS and VSS concentrations were found to be 0.9302 and 0.9258, respectively. Rate constants (k) of second-order kinetics based on TSS and VSS concentrations for first batch biodegradation and second batch biodegradation studies were found to be 0.01438 g/L and 0.0179 g/L, and 0.0159 g/L and 0.0233 g/L respectively. Based on these rate constants, it is clear that the rates and extent of biodegradation of floral waste were significantly less than that of leaf litter.

References

Analytical, O. I. (2002). "Model 1010 Wet Oxidation Total Organic Carbon Analyzer Operator's Manual." Downloaded from http://www.aimanalytical.com/Manuals/toc1020a.pdf.

APHA, AWWA and WEF (2005). Standard methods for Examination of Water and Wastewater (21st ed.). American Public Health Association.

Awasthi, M.K., Wong, J.W.C., Kumar, S., Kumar, S., Wang, Q., Wang, M., ... Zhang, Z. (2018). Biodegradation of food waste using microbial cultures producing thermostable a-amylase and cellulase under different pH and temperature. *Bioresource Technology*, **248:** 160–170. https://doi.org/10.1016/j.biortech.2017.06.160

Bhat, S.A., Singh, J. and Vig, A.P. (2017). Instrumental characterization of organic wastes for evaluation of vermicompost maturity. *Journal of Analytical Science and Technology*, **8(1):** 2. https://doi.org/10.1186/s40543-017-0112-2

Brown, M.E. and Chang, M.C.Y. (2014). Exploring bacterial lignin degradation. *Current Opinion in Chemical Biology*, 19: 1–7. https://doi.org/10.1016/j.cbpa.2013.11.015

Goel, S. (2017). Solid and Hazardous Waste Management: An Introduction. *In:* Advances in Solid and Hazardous Waste Management (1st ed.). New Delhi: Springer and Capital Books.

Goel, S., Hozalski, R.M. and Bouwer, E.J. (1995). Biodegradation of NOM: Effect of NOM source and ozone dose. *Journal/American Water Works Association*, **87(1):** 90–105. https://doi.org/10.2307/41295155

Hajjouji, H. El, Fakharedine, N., Baddi, G.A., Winterton, P., Bailly, J.R., Revel, J.C. and Hafidi, M. (2007). Treatment of olive mill waste-water by aerobic biodegradation : An analytical study using gel permeation chromatography, ultraviolet-visible and Fourier transform infrared spectroscopy. *Bioresource Technology*, **98:** 3513–3520. https://doi.org/10.1016/j.biortech.2006.11.033

Hiremath, B.R. and Goel, S. (2017). Survey of Municipal Solid Waste (MSW) Treatment Methods and Compost Samples. *In:* Advances in Solid and Hazardous Waste Management (1st ed.). New Delhi: Springer and Capital Books.

Jadhav, A.R., Chitanand, M.P. and Shete, H.G. (2013). Flower Waste Degradation Using Microbial Consortium. *IOSR Journal of Agriculture and Veterinary Science*, **3(5):** 1–4. Retrieved from www.iosrjournals.org

Jouraiphy, A., Amir, S., Gharous, M. El, Revel, J. and Hafidi, M. (2005). Chemical and spectroscopic analysis of organic matter transformation during composting of sewage sludge and green plant waste. *International Biodeterioration & Biodegradation*, **56:** 101–108. https://doi.org/10.1016/j.ibiod.2005.06.002

Kandakatla, P., Mahto, B. and Goel, S. (2013). Extent and rate of biodegradation of different organic components in municipal solid waste. *International Journal of Environment and Waste Management*, **11(4):** 350. https://doi.org/10.1504/IJEWM.2013.054262

Klotzbücher, T., Filley, T.R., Kaiser, K. and Kalbitz, K. (2011). A study of lignin degradation in leaf and needle litter using 13 C-labelled tetramethylammonium hydroxide (TMAH) thermochemolysis: Comparison with CuO oxidation and van Soest methods. *Organic Geochemistry*, **42:** 1271–1278. https://doi.org/10.1016/j.orggeochem.2011.07.007

Liu, Y., Hu, T., Wu, Z., Zeng, G., Huang, D., Shen, Y., He, X., Lai, M., and He, Y. (2014). "Study on biodegradation process of lignin by FTIR and DSC." Environmental Science and Pollution Research, 21(24), 14004–14013.

Loubna, E.F., Zamama, M., Abdelghani, E.A. and Mohamed, H. (2014). Assessment of biotransformation of organic matter during co-composting of sewage sludge-lignocellulosic waste by chemical, FTIR analyses, and phytotoxicity tests. *International Biodeterioration & Biodegradation*, **87:** 128–137. https://doi.org/10.1016/j.ibiod.2013.09.024

Tosun, I., Gonullu, M.T., Arslankaya, E. and Gunay, A. (2008). Co-composting kinetics of rose processing waste with OFMSW. *Bioresource Technology*, **99:** 6143–6149. https://doi.org/10.1016/j.biortech.2007.12.039

Worrell, W.A. and Vesilind, P.A. (2012). Solid Waste Engineering (Second). Stamford, USA: Cengage Learning. https://doi.org/10.1017/CBO9781107415324.004

Yadav, I., Juneja, S.K. and Chauhan, S. (2015). Temple Waste Utilization and Management: A Review. *International Journal of Engineering Technology Science and Research*, **2**(special): 14–19. Retrieved from www.ijetsr.com

Chapter 2
The Treatment of Landfill Leachate and Other Wastewaters Using Constructed Wetlands

Hari Bhakta Sharma, Sagarika Panigrahi, and Brajesh K. Dubey

1 Introduction

The world we are living in is slowly getting urbanized. The world's urban population which accounts for 55% today will rise to 66% by 2050 (World Urbanization Prospects: the 2014 Revision UN) (United Nations 2014). This increasing influx of the population into the cities has increased the stress in the water demand, as most of the cities are already facing severe water crisis (Hanjra and Qureshi 2010; Jury and Vaux 2007). Even if the water demand is met, it will create an additional challenge to treat wastewater generated, which accounts for almost 80% of water supplied. The latest report by United Nations World Water Assessment Programme (WWAP) (2017) highlighted that the 80% of the wastewater generated in the entire world is disposed without treatment, directly into the water bodies or open drainage and this figure is 95% in some poor countries. Disposal of this untreated wastewater in the lake, river, and stream will lead to unprecedented deterioration of health of these water bodies (Edokpayi et al. 2017; Lee et al. 2016; Khan and Ansari 2005). Untreated wastewaters have a very high concentration of nutrients, possibly heavy metals, emerging contaminants, suspended solids and pathogens, which alter the physical, chemical and biological properties of water, and thereby affect natural properties of water bodies, and biological life that thrive in it (Khan and Ansari 2005). Eutrophication, also known as an algal bloom, is one of the worst ramifications of excessive nutrient loading in water bodies due to discharge of nutrient-rich (nitrogen and phosphorus) wastewater in it (US EPA 1999; Khan and Ansari 2005).

The increase in the nutrient content in the water bodies increases the productivity of the lake, thereby increasing the growth of phytoplankton. This phytoplankton becomes the food for the algae (Huang et al. 2017; Ngatia et al. 2017; Ulloa et al.

H. B. Sharma · S. Panigrahi · B. K. Dubey (✉)
Civil Engineering Department, Indian Institute of Technology Kharagpur, Kharagpur, West Bengal, India
e-mail: bkdubey@civil.iitkgp.ac.in

D. Sengupta et al. (eds.), *Treatment and Disposal of Solid and Hazardous Wastes*,
https://doi.org/10.1007/978-3-030-29643-8_2

2017). Excessive growth of algae, fuelled by nutrient-rich wastewater covers the surface of water bodies, which is known as eutrophication, often called as an algal bloom (Anza et al. 2014; Khan and Ansari 2005; Glibert 2017). Other effects like pathogenic activity and accumulation of heavy metal and emerging pollutants have a direct implication on public health (Daley et al. 2015). The problem of treating wastewater has become even more serious in the developing nations, where availability of modern day wastewater treatment technology is in scarce (Chatterjee et al. 2016; Morel 2006). Taking into account the repercussion of untreated wastewater discharged in natural water bodies, and its effect on human health and environment, the need of decentralized treatment facility, which is less energy extensive, easily acceptable by locals, require less skill manpower and has zero carbon footprint, has become a necessity (Zaharia 2017).

When water flows through river, lake and streams, it gets purified naturally and one of these purifying processes is filtration (Cronk and Fennessy 2016; Gopal 1999; Idris et al. 2014). The other purification process that takes place in natural water bodies are sedimentation, biological assimilation of nutrients by plants and microorganism, adsorption etc. Natural wetland is one such body, where purification is relatively faster due to its complex composition of plant, soil and microorganism. Wetland soil could be always saturated with water or gets saturated occasionally. Availability of water determines the kind of plants it grows and kind of microorganism that thrives. A natural wetland is one among other several natural systems that were included into the definition of wetland by Ramsar Convention on Wetland of International Importance in 1971. Ramsar convention is an intergovernmental treaty for sustainable and wise use of natural wetland, which came into force in 1974, 4 years after it was adopted. It provides framework for conservation of wetland and its resources. A function of wetland is advantageous to both human and wildlife, due to its rich biodiversity (Knight 1997). Turbulent water becomes calmer as it enters wetland due to its interaction with vegetation; as a result suspended particle gets time to settle down and becomes a part of wetland sediment. Other pollutants are transformed to less soluble forms and some are taken up by the plants. The root matrix of wetland plants where biofilms get formed creates necessary conditions for microorganism to live and thrive on (Brix et al. 2002; Tu et al. 2014). A series of complex biochemical processes by these microorganisms makes transformation and removal of pollutants from wastewater possible (Brix et al. 2002; Stottmeister et al. 2003). The nitrogen and phosphorous which are major nutrients found in stormwater runoff, agricultural runoff and landfill leachate get deposited in wetland sediment, and are often absorbed by wetland soils and assimilated by plants and microorganisms (Koskiaho et al. 2003).

In the last few decades, natural wetland processes have been engineered in smaller scale for water quality improvement (Vymazal 2002, 2010a, b). This smaller, engineered version of wetland is known as constructed wetland (CW) (Brix 1993). The CW is an engineered system, designed to mimic natural treatment processes in the close conglomeration of wetland plants, sediment, and microorganism to improve water quality. Ever since West Germany built the first CW in 1974 (Zhang and Weiming 2011), different kinds of CW have been created to treat wastewater around the world (Hernandez-Crespo et al. 2017; Liu et al. 2017; Wang et al. 2017). The CW can be designed to treat different types of wastewater cost effectively with minimum usage of energy and manpower (Gazea et al. 1996; Liu et al. 2017; Wang et al. 2017). The CW can also be designed to remove nitrogen, phosphorous, heavy metals, solids, organic matter and pathogens from the wastewater (Vymazal 2013; Konnerup et al. 2009; Kivaisi 2001).

For underdeveloped countries of Asia and Africa treating wastewater by using conventional sewage treatment plants has always been a costly affair. These countries have neither the resource to meet energy demand much needed for treatment plants nor have skilled manpower to run the plant, except in the capital and big cities (Chatterjee et al. 2016). The minimum use of energy, no use of chemicals, the establishments of ecological habitat, aesthetically appealing are some of the added advantages of choosing CW as a wastewater treatment system which makes CW an epitome of sustainable approach, where science and nature go hand in hand. Thus integrating the wetland into treatment infrastructure will unleash its potential without affecting its basic utility as an ecosystem.

2 Introduction to the Constructed Wetland

The CW is an engineered system of water bodies designed as per need to treat pollutants found in sewage, industrial effluent or stormwater runoff (Jiang et al. 2016; Guo et al. 2017). The constructed wetlands are generally used in a decentralized manner in small communities, but in recent times it is being seen as a treatment alternative attributed largely to its evolution for treating few of the emerging pollutants (Sgroi et al. 2017; Matamoros et al. 2017; Dan et al. 2017). Vegetation (roots, stems and leaves) in a CW acts as a substrate upon which microorganisms can grow and develop biofilms as they breakdown organic materials (Kadlec and Wallace 2009; Shelef et al. 2013; Brix 1997). The community of microorganisms that thrive on the soil and root matrix of wetland is known as the periphyton (Liu et al. 2016). The biochemical activities by periphyton and natural physico-chemical processes are responsible for approximately 90 percent of pollution removal and breakdown (Tilley et al. 2014). The plants remove about 7–10% of pollutant by assimilation. It also acts as a carbon source for the microbes when they decay. Different species of aquatic plants have different rates of heavy metal uptake capacity, so knowledge of same is must while designing the CW for removing specific pollutants.

2.1 Types of Constructed Wetland

A CW can be categorized into two major groups according to the flow regime of wastewater that passes through it (Vymazal 2001; Brix 1997). These major groups are: surface flow CW and subsurface flow constructed wetland. The subsurface flow CW is further categorized as horizontal flow CW, vertical flow CW and hybrid flow constructed wetlands. In the free surface flow CW, the water flows freely above the ground, in a saturated media. The floating vegetation or emigrant plants are grown in this type of CW. Baffles are also used to control the flow of wastewater flowing through it, which gives more opportunity for a suspended matter to settle. Filtration and adsorption by wetland media, oxidation, reduction and precipitation by physicochemical processes catalyzed by the internal and external agent (sunlight, pH etc.) are the few mechanism by which the pollutants get removed from wastewater when it passes through the CW (US EPA 2000). It is very important to note that the wastewater should undergo preliminary treatment (sedimentation) before it is treated with CW. This preliminary treatment is done to minimize the clogging of filter media by settable solids. The plant root releases oxygen into its vicinity of its roots hair, which creates a condition necessary for the diverse biological and chemical processes to take place (Tilley et al. 2014).

Unlike in surface flow, where wastewater flows above the saturated media, in a subsurface flow CW, wastewater flows below the surface, within the media. The media are made up of sand or gravel. The media is kept unsaturated periodically. The subsurface flow CW is further classified into horizontal and vertical flow CW depending on the direction of flow of wastewater.

Figure 2.1 presents the different type of CW based on the flow direction. The direction of arrow tells us the direction of wastewater that flow in a CW. Based on

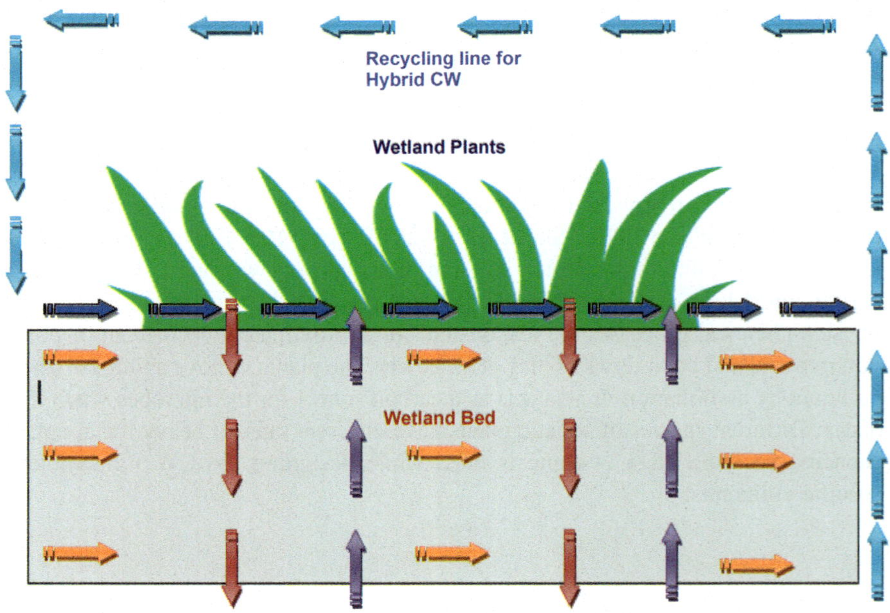

Fig. 2.1 Schematic presentation of different type of constructed wetland (CW) based on types of flow: Horizontal flow CW = ; Down flow vertical CW = ; Up flow CW = ; Surface flow CW = ; Recycling line for hybrid CW =

this flow, the CW is named as surface flow, subsurface flow, vertical flow or hybrid flow constructed wetland. As the name suggests, horizontal flow CW is one in which the flow of wastewater is horizontally in the subsurface. The media generally used are sand and gravel. Generally, the media is placed in well graded fashion with the large size gravel placed at the bottom to facilitate the drainage layer. In the top layer, the wetland plant is grown, roots of which grow to the layers below as well, making dense root matrix needed for effective pollutant removal.

The sand-gravel media not only act as a media for plant to grow on, but also act as the filter media for straining the solid particles in the voids it forms during interlocking. The media further provides the surface, upon which the bacteria can get attached to form bio-slime layers when organic-rich wastewater washes it. The root also transfers small amount of oxygen to its vicinity, thereby allowing the aerobic bacteria to colonize and degrade the organic matter. The permeability of the filter is maintained well by the plant's roots, generally through the area of root media interface (Tilley et al. 2014).

In the vertical flow constructed wetland, the flow of water is kept vertical to the subsurface layer. The flow is either upward or downward. The wastewater is intermittently applied throughout the bed. The filter undergoes the phases of the saturated and unsaturated condition during loading and non-loading of wastewater.

The wastewater percolates down the unsaturated bed, as the bed drains; air is also drawn along with it. The porous media give sufficient time for oxygen to diffuse throughout the media. The organic matter can be degraded by the aerobic bacteria near the root zone, where the oxygen is transferred by root. The deep root growth maintains the permeability of filter media. To decrease the excess biomass growth and to increase the porosity of media dosing rate and time is varied which forces the microorganism to starve (Hoffmann et al. 2011). To increase the treatment efficiency, different types of CW are linked to form a hybrid system. The water flowing out (effluent) of one wetland will be the influent for the second wetland in a series for a hybrid CW. The hybrid system reduces total nitrogen concentration from wastewater by enhancing nitrification and denitrification (USEPA 2000).

2.2 *Plants Used in Constructed Wetland*

The plants used in the CW can be broadly classified into three groups (Brix 2003): (1) Free-floating plants (e.g. *Eichhorniacrassipes, Lemna, Spirodella, Wolffia*); (2) Emergent plants (e.g. *Phragmites, Scirpus, Typha, canna*) and (3) Submerged plants (e.g. *Isoteslacustris, Elodea canadensis*). It should be understood that all the type of plant, namely, floating, submerged and emergent, all help in increasing oxygen concentration within the system thereby provide necessary biological as well as the chemical condition for degradation of pollutants. The floating plants derive nutrient from wastewater but very less from the substrate, unlike emergent and submerged which derive nutrient mostly through substrate. The nutrient gets adsorbs in the filter media which the microorganism and plant can assimilate upon.

The submerged and emergent plants provide a maximum surface area for biofilms growth; they are also better in increasing sedimentation rate of solid as their roots and stem hinder the flow rate of incoming wastewater.

2.3 *Role of Plant Macrophytes in Constructed Wetland*

The performance of CW largely depends on the root matrix created by wetland. The hollow system in the plant tissues enables oxygen to be transported from the leaves to the root zone and to the surrounding soil, which helps in creating aerobic condition in its premises. Figure 2.2 illustrates the role of the plant in CW. This diffused oxygen by plant facilitates aerobic degradation of pollutants. The role of wetland plants in CW systems can be divided into six categories as explained below.

1. **Physical** – Plant controls the surface of media beds, helps in a filtration, and provides a large surface area for microbial growth. The growth of plants minimizes inflow velocity of wastewater which increases sedimentation rate of solids (Brix et al. 2002).
2. **Increase in porosity** – The plant root is seen to improve the porosity of wetland soil media through root media interface. The increase in porosity improves

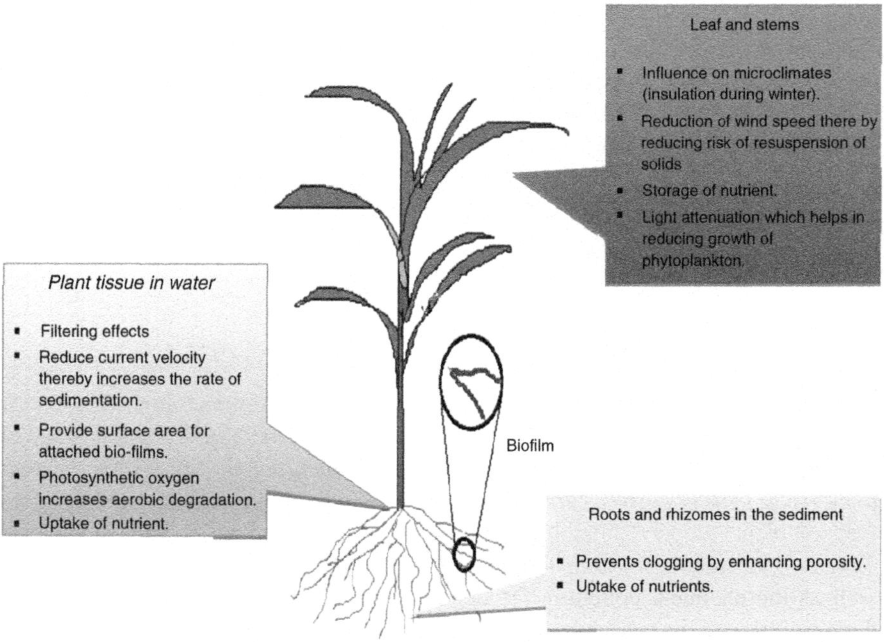

Fig. 2.2 The role of plants in the constructed wetlands (Brix 1997)

aerobic condition in the system which helps in aerobic degradation of pollutants (Brix et al. 2002).

3. **Microbial growth** – Macrophytes provide a large surface area for growth of microbial biofilms. These biofilms are responsible for a majority of the microbial processes in a CW, including nitrogen removal (Brix et al. 2002).
4. **Creation of aerobic soils** – Macrophytes facilitates the transfer of oxygen through the pores of plant tissue and transfers it to rhizosphere via root system where aerobic removal of organic matter and nitrification takes place (Reddy et al. 1990; Brix et al. 2002).
5. **Aesthetic values** – The wetland plants provide a habitat for wildlife and makes wastewater treatment systems aesthetically and publicly acceptable.

Flower bearing wetland plants like canna lilies increases the aesthetic value which is easily accepted by public (Ojoawo et al. 2015; Brix et al. 2002). The wetland premise can be open up for public which helps in generating revenue.

3 Landfill Leachate Treatment Using Constructed Wetland

To better understand the treatment of landfill leachate using CW, it becomes imperative to understand characteristics of landfill leachate. Landfill leachate is a complex polluted liquid that gets formed due to passing of rainwater through landfill and brings along out with it high amount of dissolved as well as suspended matter. The formation of the dissolved and suspended matter is due to biochemical process that takes place inside the landfill. The landfill leachate is characterized by high amount of organics (BOD and COD), nutrient (nitrogen and phosphorus), dissolved solids, heavy metal etc. The characteristics of typical landfill leachate are presented in Table 2.1. However, the composition of landfill leachate varies significantly with respect to landfill age (Renou et al. 2008). The main mechanisms through which the contaminants are leached through the landfill are dissolution of soluble materials, biodegradation of organic matters, chemical reduction and washing of fine materials (Bricken 2003).

Section below explains in detail about contaminant from landfill leachate gets removed in a CW. Removal mechanism of nutrient like nitrogen and phosphorus is explained in detail.

3.1 Organic Matter Removal

The removal efficiency of the organics (BOD_5 and COD) in the CW is very high. Up to 90–99% removal efficiency of BOD_5 and COD is reported in literature (Vymazal et al. 1998; Kadlec and Knight 1996). Under normal condition, the settleable organic is removed by simple sedimentation and filtration. Both anaerobic and aerobic activity

Table 2.1 Characteristics of raw landfill leachate

Parameters	Units	Ghafari et al. (2009)	Al-Hamadani et al. (2011)	Othman et al. (2010)	Akinbil et al. (2012)	Nivala et al. (2007)	Bulc (2006)	Yalcuk and Ugurlu (2009)
pH		8.4	8.13	8.26	8.42	–	–	–
Colour	TCU	3869	3140	2933	3360	–	–	–
Turbidity	NTU	–	–	–	140	–	–	–
TSS	mg/L	80	380	130	685	186	38.3	–
COD	mg/L	1925	2130	3180	923.44	781	485	4770
$BOD_5^{20\ °C}$	mg/L	–	–	231	686	116	76	–
NH_3-N	mg/L	1184	1950	2050	238	212	496	2865
Phosphorus (P)	mg/L	–	–	–	117	–	2.3	75
Nitrogen (N)	mg/L	–	–	–	400	–	–	67.7
Iron	mg/L	–	–	5.3	6.19	21	–	–
Manganese	mg/L	–	–	0.23	24.8	–	–	–
Magnesium	mg/L	–	–	–	660	–	–	–
Zinc	mg/L	–	–	3.4	7.43	–	–	–

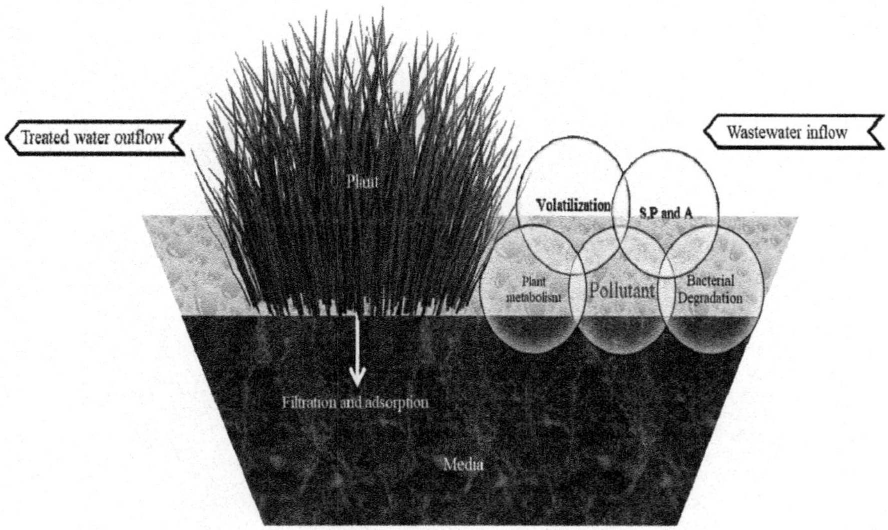

Fig. 2.3 Fate of pollutant in the constructed wetland (S, P&A = Sedimentation, Precipitation and Adsorption)

takes place to degrade the organics in wetlands. Figure 2.3 illustrates the fate of pollutants in the constructed wetland. The degradation of organics takes place in the biofilms established by microorganism around the filter media, roots etc. (Kadlec 2009).

The oxygen required for the process to undergo is met through direct diffusion, also called as surface transfer, or by oxygen leakage from the root of macrophytes. The root system that gets developed deeper into the lower media layer will maximize the pollutant removal in all parts of media (Tilley et al. 2014). The deep root system which increases the porosity of media also enhances the movement of oxygen diffused through wastewater from surface, thus enhancing oxygen availability (Hoffmann et al. 2011).

3.2 Phosphorous Removal

Phosphorus is present in wastewater as orthophosphate, polyphosphate and organic phosphorus. The conversion of most phosphorus to the orthophosphate forms is caused by biological oxidation (Vymazal et al. 1998). The phosphorous removal is achieved by adsorption, assimilation and precipitation reaction involving iron and aluminum; so it can be said that the removal of phosphorus depends on the concentration of iron and aluminum in the filter media. The phosphorous is removed through precipitation by iron (Fe) and aluminum (Al) (Vymazal et al. 1998). It is to be understood that only ortho phosphorous is believed to be directly utilized by algae and macrophytes (Vymazal 1995).

Fig. 2.4 Fate of phosphorous in the constructed wetland

Uptake by macrophytes is accounted to be low, as it removes only small fraction of the total phosphorus. The harvesting of biomass to enhance the uptake of phosphorus by plants and thus removal is recommended (Vymazal et al. 1998; Wang et al. 2015). The fate of phosphorus in the constructed wetland is illustrated in Fig. 2.4. It is estimated that a phosphorus removal ratio by plant growth of up to 10% is possible depending on the climate, plant and type of wastewater. The phosphorous removal efficiency decreases over time due to the decrease in the adsorption sites in the filters (sand and gravel). To overcome this problem, a separate unplanted filter media is advised to be constructed in downstream to enhance phosphorous precipitation (Hoffmann et al. 2011) and to keep treatment system running.

Phosphorus is an essential requirement for biological growth. An excess of phosphorus can have secondary effect of triggering eutrophication within a wetland which is caused by an algal blooms. The phosphorous removal in all types of CW is low but this can be increased by using proper substrate, the substrate with high Al and Fe content (Vymazal 2007). The red soil has high iron content which has a potential to reduce the phosphorous in the wastewater by forming the phosphorus complex. The high phosphorous removal efficiency in the vertical CW with red soil is observed (Villar et al. 2012).

3.3 Nitrogen Removal

Nitrogen concentration in wastewater is often of concern because of its potential to cause adverse effect in receiving water systems. Among various nitrogen groups, dissolved inorganic nitrogen species like nitrate (NO_3), nitrite (NO_2), and ammonia (NH_3) or ammonium (NH_4) have the greatest impact on aquatic systems, because they are easily available for uptake by microorganisms. The removal mechanisms for nitrogen in constructed wetlands include ammonification, nitrification/denitrification, plant uptake and root system adsorption (Chang-gyun et al. 2009; Korkusuz et al. 2004). The removal of organic substances, typically 80–90%, is achieved in constructed wetlands. However, the nitrogen removal rates are often not up to the mark. A variety of nitrogen forms in constructed wetlands can be removed through specific treatment processes, such as combined nitrification-denitrification and sedimentation, particularly at the sediment-water and water-plant interface (Vymazal et al. 1998; Lee et al. 2009). Figure 2.5 elucidates the fate of nitrogen in CW.

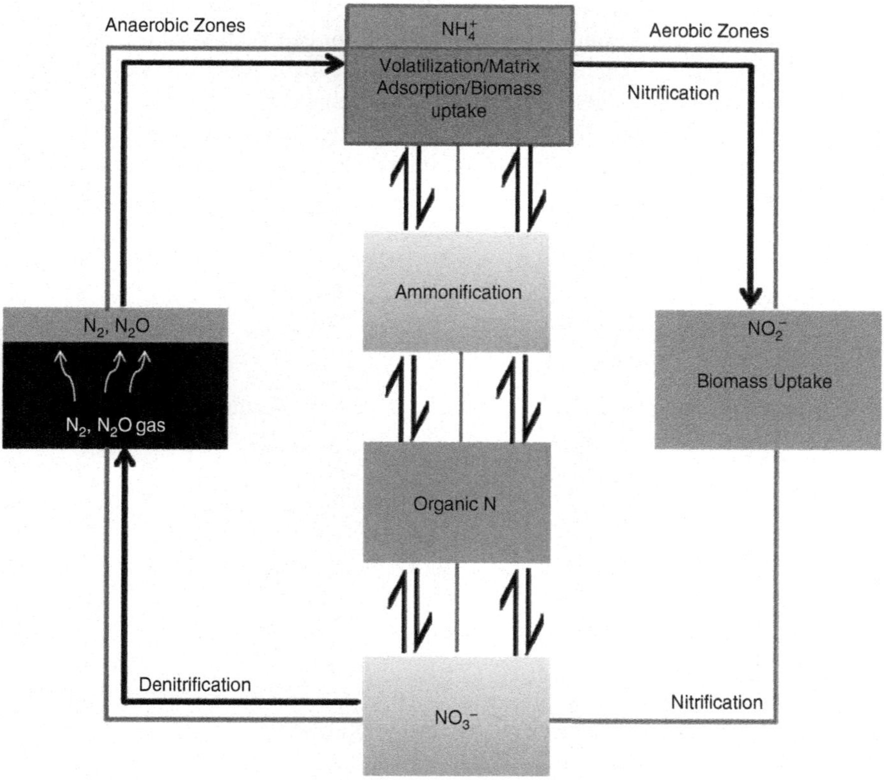

Fig. 2.5 Nitrogen fate in the constructed wetland

With proper aeration facility the total nitrogen removal efficiency can be increased up to 90% or even greater (Davies and Hart 1990). Vymazal (2007) found out that the total nitrogen removal efficiency in his established constructed wetlands varied between 40% and 50%, depending on CW type and inflow loading. It was inferred that single stage treatment wetland is less efficient than the hybrid system due to inability to provide both the conditions of aerobic and anaerobic needed for nitrification and denitrification to take place. It was further quoted that vertical flow CW successfully removes ammonia but very little denitrification takes place and on the contrary, the horizontal flow CW provides good condition for denitrification but its ability is limited for nitrification only (Vymazal and Kröpfelová 2011). Wang et al. (2009) found out that, with increasing hydraulic load, the removal efficiency of COD was stable. The wetland can resist the impact of COD but the removal rate of TN (total nitrogen) decrease with the increase in hydraulic and pollution concentration loading rate. The increase in hydraulic loading rate and decrease in oxygen release by plant in the wetland can be the reason for less TN removal (Wang et al. 2009). The C:N (carbon-nitrogen ratio) is also the reason for less removal of TN in the constructed wetland. The TN removal happens best in the high ratio of carbon and nitrogen. Both of the bacteria nitrosomonas and nitrobacteria use the inorganic carbon compound for cell synthesis. The denitrification is the process by which nitrite and nitrate are eventually converted to nitrogen gas while oxidizing the carbon for energy (Khanitchaidecha et al. 2010; Carley and Mavinic 1991). Nitrogen can also get adsorbed to sand and soil differently, the substrate of the sands can absorb N to a certain degree, which has a significant effect in TN removal (Zhang et al. 2007).

3.4 *Nitrification*

Nitrification is a two-step process, where conversion of ammonia to nitrate takes place in strict aerobic condition.

(a) Ammonia to nitrite

$$(NH_4{}^+ + 1.5O_2 \rightarrow NO^-{}_2 + H_2O + 2H^+)$$

(b) Nitrite to nitrate

$$(NO^-{}_2 + 0.5O_2 \rightarrow NO^-{}_3)$$

Vymazal et al. (1998) pointed out that the nitrification is influenced by temperature and pH. The intermittently loaded vertical flow CW have the high rate of oxygenation; therefore it provides good nitrification (Brix 1998). The oxidation reactions release energy used by both nitrosomonas and nitrobacter cell synthesis. The combined processes of cell synthesis create 0.17 g of dry weight biomass per gram of ammonia nitrogen consumed (Kadlec and Wallace 2009). Nitrification is an aerobic process carried out by autotrophic bacteria, commonly nitrosomonas and nitrobacteria. They derive energy from the oxidation of ammonia and nitrite. The nitrosomonas only oxidize ammonia to nitrite while later can only oxidize nitrite to nitrate. Both of the bacteria use the inorganic carbon compound for cell synthesis. The growth rate for nitrosomonas is less than that of nitrobacter, thus nitrite accumulation should not occur in the aerobic basin unless nitrobacter is inhibited (Khanitchaidecha et al. 2010).

3.5 *Denitrification*

The first oxidation process to occur after oxygen depletion is the reduction of nitrate to molecular nitrogen and nitrogen gases, which is called as denitrification. In a vertical-flow constructed wetlands very high nitrification process takes place because of entirely aerobic conditions but no denitrification takes place. In order to achieve effective removal of total nitrogen, vertical flow CW should be combined with horizontal flow CW which provides suitable conditions for reduction of nitrate formed during nitrification to free nitrogen (Brix and Arias 2005). Denitrification is the process by which nitrite and nitrate are eventually converted to nitrogen gas while oxidizing the carbon for energy. The denitrifying bacteria are *Pseudomonas*, *Micrococcus*, and *Bacillus* (Khanitchaidecha et al. 2010).

3.6 *Pathogen Removal*

The public health is always the centre of reason for treating wastewater. All the treatment methods try to address this centre reason (Metcalf and Eddy 1991). Pathogens are seen predominantly in domestic wastewater. The conventional treatment process adds up some chemicals like chlorine and ozone to get rid of this pathogen. Also, the higher operational and maintenance cost limits the use of ozonation and ultraviolet disinfection for pathogen removals (Metcalf and Eddy 1991). The CW is better known to eliminate these constraints with its suitable combination of a physical, chemical and biological factor for pathogen removal (Vymazal 2005). The pathogen treatment in wetland systems relies on sedimentation, natural die-off, temperature, oxidation, predation and UV radiation (Vymazal 2005; Alufasi et al. 2017). The pathogen removal is most often required when dealing with domestic wastewater. The reduced microbial activity, reduced plant activity, and freezing of the water can all occur due to cold temperature. The human pathogens function most efficiently around internal body temperatures (~37 °C). It is found out that decreased temperatures outside will inactivate pathogens. The inactive pathogens are easy to treat (Weber and Legge 2008). Some bacteria are facultative or anaerobic and thus the presence of oxygen creates unfavourable conditions for these organisms (Vymazal 2005).

Song et al. (2008) investigated the efficiency to remove pathogen by the construal wetland and found out that the system effectively removes *E. Coli*, fecal and total coliforms. Zhang and Weiming (2011) have investigated the effect of water temperature, water quantity, water loads, influent concentration and pH in the treatment efficiency of the constructed wetland. These five impact factors selected were used in PCA (principal component analysis) method which helps to establish relation between the removal efficiency and the impact factors. The analysis was done for total nitrogen (TN), total phosphorous, BOD_5, suspended solids and fecal coliform. The average removable rate of each was greater than 60% and for fecal coliform it was 99%. It was also found that the temperature, concentration, and load have significant impact than the pH and water quantity. It was found out that the temperature plays vital role in microbe's activity.

4 Constructed Wetland for Treatment of Different Types of Waste Water

4.1 CW for Removal of Emerging Contaminant

Emerging contaminants are called so as they are new groups of compound with little to nil knowledge about its eco-toxicological effect. This includes pharmaceutical products, personal care products, surfactants, plasticizers, herbicides etc. (Murray et al. 2010). The antibiotic resistance gene (ARG) is also labeled as the emergent contaminant (Pruden et al. 2006). The CW tends to remove emerging pollutants like diclofenac, ketoprofen, caffeine etc. with higher efficiency as compared to a pond (Matamoros and Bayona 2006; Matamoros and Salvadó 2012). The reason of better functioning of CWs for removal of emerging contaminants is attributed to higher HRT (Hydraulic Retention Rate) and presence of plants. The specific condition which prevails in a wetland matrix enhances the removal of these emerging contaminants. For example, Avila et al. (2013) found out that the ibuprofen biodegradation takes place under aerobic condition in their horizontal flow CW. Better removal of emerging contaminants from CW is also credited for high temperature and different physico-chemical condition prevailing in the wetland which facilitates different abiotic/biotic factors for its effective degradation. These different factors are biodegradation, sorption, volatilization, hydrolysis, and photo-degradation (Avila et al. 2015). Table 2.2 highlights few emerging contaminants removed by the different type of CW and its removal mechanism. The study by Chen et al. (2015) found that to remove antibiotics and antibiotic resistance gene from wastewater using CW, the major mechanism for removal of antibiotics in CW is due to adsorption onto media and sludge and very little is attributed to biodegradation.

Table 2.2 Few emerging contaminants removed (percentage removal) by different type of constructed wetlands and its removal processes

Some emerging contaminants	Diclofenac	Kteoprofen	Ibuprofen	Salicylic acid	Caffeine	Naproxen	Types of CW	References
Percentage removal	<50	>85	50–85	>85	50–85	50–85	SSFCW	Zhang et al. (2012)
	<50	<50	50–85	>85	>80	50–85	SSFCW	Matamoros and Bayona (2006)
	55–95	–	>98	–	–	–	VFCW	Avila et al. (2014)
	89	–	99	–	–	–	HFCW	Avila et al. (2015)
	19–21	–	12–44	–	–	57–75	HFCW	Matamoros et al. (2017)
	70–90	>95	>95	>95	70–90	–	VFCW	Matamoros et al. (2007)
Removal mechanism of different emergent contaminants	Bio degradation, photo degradation and plant uptake[a]	Temperature dependent bio-degradation[b]	Aerobic degradation[c]	–	Anaerobic biodegradation and plant uptake[d]	Plant uptake[e]		[a] Matamoros et al. (2012), [b] Zhang et al. (2012), [c]Hijosa-valsro et al. (2010a, b), [d] Zhang et al. (2013), and [e]Hijosa-valsero et al. (2010a, 2011)

Note: SSFCW = Sub surface flow constructed wetland; VFCW = Vertical flow constructed wetland; HFCW = Hybrid flow constructed wetland

Table 2.3 Domestic wastewater treatment by constructed wetland (Influent and effluent concentration)

	Influent, mg/L	Effluent, mg/L	References
BOD_5	100–400	0–12	Shrestha et al. (2001)
	392	81	Karathanasis et al. (2003)
COD	177–687	7–72	Shrestha et al. (2001)
Nitrogen	4–26	0–2	Shrestha et al. (2001)
	1681	635	Nakamura et al. (2017)
Phosphorous	1–5	1–4	Shrestha et al. (2001)
	561	15	Nakamura et al. (2017)

The photo-degradation also contributes to the abiotic transformation of the emergent contaminant in the wetland; this transformation is however limited by pH and dissolved organic carbon (Jasper and Sedlak 2013).

The major antibiotics removed in the study by Chen et al. (2015) were ofloxacin, anhydrous-erythromycin and sulfamethazine. The ARGs removal is mainly attributed to biodegradation and sorption onto media. Chen et al. (2016) in his studies found out that mesocosm scale CW can remove antibiotic and ARG on par with conventional wastewater treatment systems. The study further suggested that using different filter media, this removal efficiency can be further enhanced.

4.2 Constructed Wetland for Sewage Treatment

Since the water demand in the urban area is very high and if the cheap treatment alternative provides reusable water, the stress in water demand will be greatly reduced. The constructed wetland can treat the domestic wastewater with high removal of BOD_5, nitrogen, phosphorous etc. (Liu Wen et al. 2011) and make water fit for reuse for various purposes (Vymazal 2010a, b).

Table 2.3 highlights removal of few organic contaminants by CW. Shibao Lu et al. (2015) constructed a multi-layer CW for treatment of domestic wastewater for research purpose. The wetland was evaluated for the removal of COD, BOD_5, nitrogen and phosphorous. The results were positive and the removal rates were 90.6%, 87.6%, 66.7% and 90% respectively.

4.3 Constructed Wetland for Industrial Wastewater Treatment

The alternative treatment system like CW faces a lot of question when it is used for industrial wastewater treatment. Most of the industries use the conventional mechanical and chemical based treatment process. For any process to replace this system it should meet few prerequisite like fast and easy treatment process and low area requirements. However, the need for industries to implement green norms can be a great driving factor to accept CWs as a treatment infrastructure which helps them maintain their green image (Wallace 2010). Ongoing research related to the CW focus mainly on industries specific contaminant removal unique to the particular industry. Some of the applications of CW in the industrial wastewater treatment are:

- Leachate control from landfill (Yi et al. 2017; Mojiri et al. 2016; Madera-Parrra et al. 2015).
- Mines drainage (Mitsch and Wise 1998; Gandy et al. 2016)
- Pulp and paper industry (Choudhary et al. 2001; Arivoli et al. 2015)

Zhang et al. (2010) studied metal uptake test by Vallisneriaspiralis in a constructed wetland. Chromium was effectively removed by adsorption and absorption into plant tissue but this drastically reduced the photosynthetic activity in the plant. Further, eight laboratory-scales CW were set up to remove metals in mine wastewater containing lead (Pb) and zinc (Zn). It was found that the wetland system can remove 90% Pb and 72% Zn. The harvest of biomass is very necessary to increase metal uptake, as new plant tissue that grow after harvest will take up metal more quickly.

4.4 Constructed Wetland for Agro-dairy Industries

Kadlec and Wallace (2009) found out that the constructed wetland treatments are compatible with typical farm and ranch operations. Types of livestock wastewater being treated by constructed wetlands include dairy manure and milk-house wash water, runoff from concentrated cattle feeding operations, poultry manure and swine manure (Cronk 1996; Julie et al. 1995). The most important constituents in animal wastewater are nitrogen and phosphorus and both of these can be reduced in constructed wetlands if conditions are suitable.

Nitrogen makes its way to an animal wastewater treatment CW in either as an organic or inorganic form. Nitrate and nitrite are inorganic forms, which may be eliminated from the wetland through volatilization, plants uptake or by nitrification process. Above satisfactory nitrogen removal rate have been found in many wetland studies as the removal of nitrogen involves microbial processes. Nitrogen removal is high during the growing season when high temperatures facilitate microbial growth. In addition, uptake of nitrogen by plants occurs during the growing season only. Adsorbed phosphorus on soil particles is taken up by the plants. Adsorption is the main process for phosphorus removal in constructed wetlands. Aerobic condition favours the phosphorus adsorption, and in the anaerobic conditions adsorption of phosphorus in wetlands is found to be less than on dry soil. As phosphorus loading to a CW happens over a period of several years, the adsorption capacity decreases and the phosphorus gets released.

4.5 *Constructed Wetland for Urban Storm Water Treatment*

Kadlec and Wallace (2009) found out that the CW can be effectively used to treat urban storm water which generally contains high suspended solids (SS), suggesting that the sedimentation pond should also be constructed along with CW to reduce the loading of SS in the constructed wetland.

The CW removes many contaminants from the water like organics, suspended solids, nitrogen, phosphorous, pathogens and trace metals (Vymazal et al. 1998). The ability of the CW to remove metals like Zn, Ni, Pb, Cu etc. from urban runoff was demonstrated (Scholes et al. 1998) in south-east England. The hybrid CW coupled with sedimentation was constructed by Choi et al. (2015) to remove COD, TN, TP and heavy metals like Fe, Cu and Zn. The removal efficiency of 51–70% was observed. In a study Griffiths and Mitsch (2017) has observed the potential removal of nutrient by wetland from urban runoff in tropics condition and have seen the significant removal of nitrogen and phosphorus. The heavy metal uptake by wetland plant is very less as compared to heavy metal deposition into wetland sediment (Gill et al. 2017). On an average 86% removal of Zn and 60%, Cu was documented over a 9-year study by Gill et al. (2017) using free flow CW which was designed to treat motorway runoff. Schmitt et al. (2015) in their study found out that CWs can remove solids and COD upto 98%. The removal mechanism for total nitrogen and total phosphorus was mainly attributed to biochemical transformation. The sediment storage of micro pollutants from storm water was also observed in CWs (Schmitt et al. 2015). However due to occurrence of large amount of suspended solids in storm water, frequent clogging of CW media may arise, thereby, reducing its utility as treatment unit.

5 Integration of Constructed Wetland and Microbial Fuel Cell

In recent years, researchers have explored the potential of harvesting dual benefits from engineering CW in such a way that it treats wastewater and at the same time it also produces electricity. This dual benefit is made possible by integrating microbial fuel cell (MFC) within the strata of CW. The integration is made possible after identifying few common grounds. Both the systems (MFC and CW) largely depend on the microbial action for their performance (Doherty et al. 2015a). The MFC consists of the cathode, an anode, an external circuit and a separator. It requires anode to remain anaerobic while the cathode is exposed to oxygen and both the electrodes are separated by the separator. At either side of this separator oxidation and reduction take place (Yadav et al. 2012). Interestingly CWs provide both this type of aerobic as well and anaerobic zone, where oxidation and reduction take place (Doherty et al. 2015a). Since the ability of CW to treat wastewater is time tested and well established, a natural concern is the fact that will the integration of MFC with

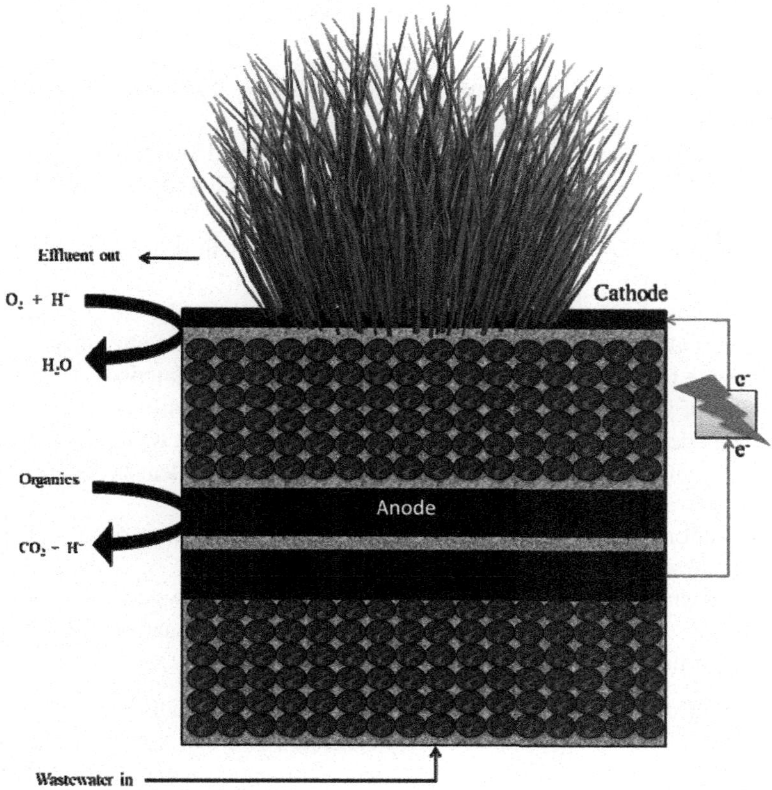

Fig. 2.6 Integration of constructed wetland and microbial fuel cell. (Adopted from Fang et al. 2017; Doherty et al. 2015a)

CW undermine the other's performance. To answer this question, researchers have found that coupling two systems actually in fact increase the performance by providing synergy effect (Fang et al. 2013; Doherty et al. 2015).

Figure 2.6 illustrates the integration of CW with the microbial fuel cell. The study by Fang et al. (2013) for decolorization of azo dye in MFC coupled CWs found out that the presence of anode significantly improved dye decolorization by 15%. The result was tested with the open circuit which shows less percentage decolorization. The dyes get adsorbed at anode surface where bio-films get developed and later get oxidized by bacteria which releases more electrons which ultimately helps in functioning of MFCs (Fang et al. 2013). Similar result was also obtained by Fang et al. (2015), where dyes decolorization varied from 36% to 66%. The integration of CWs and MFC is subject to great optimization of different parameters such as organic loading rate (COD loading significantly affects the system), redox condition (redox gradient between anode and cathode needs to be prudently optimized), wetland plants and bacteria (Doherty et al. 2015).

6 Recommendation and Conclusions

The following points highlight our suggestion for successful implementation of CW in developing nations:

- Use of free land/arid land: Using available free and arid land will bring down the cost to a large extent. Constructing CW in the arid land will also help in creating habitat for many species of amphibians and reptiles.
- Using cheap clay liner as an impermeable layer will eliminate groundwater contamination.
- More pilot scale studies and modelling studies similar to the study by Galanopoulos et al. (2013) will help us predict different attributes that may contribute to design better full scale CW with improved function and resiliency.
- Control of vector and odour: Bottom leaves of the plant will get dipped in water and gets decomposed; many dead leaves and dead plants will also decompose. Therefore it is advised to occasionally remove such fallen leaves and dead plants to minimize odour and vector problem.
- Introduction of nematodes and local fish will help in getting rid of mosquitoes larvae, thereby, reducing breeding of vectors.
- The CW premise can be upgraded as a tourist spot and environmental education centre; revenue generated from it can be used to run lab and R&D programme within wetland premise.

Constructed wetland provides beneficial treatment alternatives in many of the developing nations owing to its cost-effectiveness. The urban wetland system which is currently being researched and implemented in many countries can not only provide pleasing aesthetic appeal but also helps in treating storm-water runoff and other types of wastewater. One of the main drawbacks which arise while planning to adopt CW is the larger land requirement. For developing nations with sparse population, land requirement may not pose serious problem but, for many developed and developing nation with dense population, making land available is nearly impossible. As a result they opt for costly, modern day treatment facilities. Modern-day wastewater treatment facility comes with the high capital cost and requires skilled labour. However, developing nation lack both skill manpower as well as capitals cost.

References

Akinbil, C.O., Yusoff, M.S. and Zuki, A.A. (2012). Landfill leachate treatment using sub-surface flow constructed wetland by Cyperus haspan. *Waste Management*, **32(7):** 1387–1393.

Al-Hamadani, Y.A., Yusoff, M.S., Umar, M., Bashir, M.J. and Adlan, M.N. (2011). Application of psyllium husk as coagulant and coagulant aid in semi-aerobic landfill leachate treatment. *Journal of Hazardous Materials*, **190(1–3):** 582–587.

Alufasi, R., Gere, J., Chakauya, E., Lebea, P., Parawira, W. and Chingwaru, W. (2017). Mechanisms of pathogen removal by macrophytes in constructed wetlands. *Environmental Technology Reviews*, **6(1):** 135–144.

Anza, I., Vidal, D., Laguna, C., Díaz-Sánchez, S., Sánchez, S., Chicote, Á., and Mateo, R. (2014). Eutrophication and bacterial pathogens as risk factors for avian botulism outbreaks in wetlands receiving effluents from urban wastewater treatment plants. *Applied and Environmental Microbiology*, **80(14)**, 4251–4259.

Arivoli, A., Mohanraj, R. and Seenivasan, R. (2015). Application of vertical flow constructed wetland in treatment of heavy metals from pulp and paper industry wastewater. *Environmental Science and Pollution Research*, **22(17):** 13336–13343.

Ávila, C., Bayona, J.M., Martín, I., Salas, J.J. and García, J. (2015). Emerging organic contaminant removal in a full-scale hybrid constructed wetland system for wastewater treatment and reuse. *Ecological Engineering*, **80:** 108–116.

Ávila, C., Nivala, J., Olsson, L., Kassa, K., Headley, T., Mueller, R.A. and García, J. (2014). Emerging organic contaminants in vertical subsurface flow constructed wetlands: Influence of media size, loading frequency and use of active aeration. *Science of the Total Environment*, **494:** 211–217.

Ávila, C., Reyes, C., Bayona, J.M. and García, J. (2013). Emerging organic contaminant removal depending on primary treatment and operational strategy in horizontal subsurface flow constructed wetlands: Influence of redox. *Water Research*, **47(1):** 315–325.

Bricken, E.C. (2003). Constructed Wetlands as an Appropriate Treatment of Landfill Leachate. Doctoral dissertation, University of Natal, Durban.

Brix, H. (1993). Wastewater treatment in constructed wetlands: System design, removal processes, and treatment performance. Constructed wetlands for water quality improvement, 9–22.

Brix, H. (1997). Do macrophytes play a role in constructed treatment wetlands? Water Science and Technology, 35(5), 11–17.

Brix, H. (2003). Plants used in constructed wetland and their functions. Wetland Plants Manual, Department of Plant Ecology, Institute of Biological Science, University of Aarchus, Denmark.

Brix, H. and Arias, C.A. (2005). The use of vertical flow constructed wetlands for on-site treatment of domestic wastewater, New Danish guidelines. *Ecological Engineering*, **25(5):** 491–500.

Brix, H., Dyhr-Jessen, K. and Lorenzen, B. (2002). Root-zone acidity and nitrogen source affects *Typhalatifolia* L. growth and uptake kinetics of ammonium and nitrate. *Journal of Experimental Botany*, **53:** 2441–2450.

Bulc, T.G. (2006). Long term performance of a constructed wetland for landfill leachate treatment. *Ecological Engineering*, **26(4):** 365–374.

Carley, B.N. and Mavinic, D.S. (1991). The Effects of External Carbon Loading on Nitrification and Denitrification of a High-Ammonia Landfill Leachate. *Research Journal of the Water Pollution Control Federation*, **1(63):** 51–59.

Chang-gyun, Lee Tim, Fletcher, D. and Guangzhi Sun (2009). Nitrogen removal in constructed wetland systems. Department of Civil Engineering, Monash University. Victoria, Australia Eng. Life Sci.

Chatterjee, P., Ghangrekar, M.M. and Rao, S. (2016). Low efficiency of sewage treatment plants due to unskilled operations in India. *Environmental Chemistry Letters*, **14(3):** 407–416.

Chen, J., Liu, Y.S, Su, H.C., Ying, G.G., Liu, F., Liu, S.S. and Chen, F.R. (2015). Removal of antibiotics and antibiotic resistance genes in rural wastewater by an integrated constructed wetland. *Environmental Science and Pollution Research*, **22(3):** 1794–1803.

Chen, J., Wei, X.D., Liu, Y.S., Ying, G.G., Liu, S.S., He, L.Y. and Yang, Y.Q. (2016). Removal of antibiotics and antibiotic resistance genes from domestic sewage by constructed wetlands: Optimization of wetland substrates and hydraulic loading. *Science of the Total Environment*, **565:** 240–248.

Choi, J., Geronimo, F.K.F., Maniquiz-Redillas, M.C., Kang, M.J. and Kim, L.H. (2015). Evaluation of a hybrid constructed wetland system for treating urban stormwater runoff. *Desalination and Water Treatment*, **53(11):** 3104–3110.

Choudhary, A.K., Kumar, S. and Sharma, C. (2001). Constructed wetlands: An option for pulp and paper mill wastewater treatment. *Electronic Journal of Environmental, Agricultural and Food Chemistry* (EJEAFChe). **10(10):** 3023–3037.

Cronk. J.K. and Fennessy, M.S. (2016). Wetland plants: Biology and Ecology. CRC Press.

Cronk, J.K. (1996). Constructed wetlands to treat wastewater from dairy and swine operations: A review. *Agriculture, Ecosystems & Environment*, **58(2–3):** 97–114.

Daley, K., Castleden, H., Jamieson, R., Furgal, C. and Ell, L. (2015). Water systems, sanitation, and public health risks in remote communities: Inuit resident perspectives from the Canadian Arctic. *Social Science & Medicine*, **135:** 124–132.

Dan, A., Fujii, D., Soda, S., Machimura, T. and Ike, M. (2017). Removal of phenol, bisphenol A, and 4-tert-butylphenol from synthetic landfill leachate by vertical flow constructed wetlands. *Science of the Total Environment*, **578:** 566–576.

Davies, T.H. and Hart, B.T. (1990). Use of aeration to promote nitrification in reed beds treating wastewater. In *Constructed Wetlands in Water Pollution Control* (pp. 77–84). Pergamon.

Doherty, L., Zhao, Y., Zhao, X. and Wang, W. (2015). Nutrient and organics removal from swine slurry with simultaneous electricity generation in an alum sludge-based constructed wetland incorporating microbial fuel cell technology. *Chemical Engineering Journal*, **266:** 74–81.

Doherty, L., Zhao, Y., Zhao, X., Hu, Y., Hao, X., Xu, L. and Liu, R. (2015a). A review of a recently emerged technology: Constructed wetland–Microbial fuel cells. *Water Research*, **85:** 38–45.

Edokpayi, J.N., Odiyo, J.O. and Durowoju, O.S. (2017). Impact of Wastewater on Surface Water Quality in Developing Countries: A Case Study of South Africa. *In:* Water Quality. InTech.

Fang, Z., Song, H.L., Cang, N. and Li, X.N. (2013). Performance of microbial fuel cell coupled constructed wetland system for decolorization of azo dye and bioelectricity generation. *Bioresource Technology*, **144:** 165–171.

Fang, Z., Cheng, S., Wang, H., Cao, X. and Li, X. (2017). Feasibility study of simultaneous azo dye decolorization and bioelectricity generation by microbial fuel cell-coupled constructed wetland: Substrate effects. *RSC Advances*, **7(27):** 16542–16552.

Fang, Z., Song, H.L., Cang, N. and Li, X.N. (2015). Electricity production from Azo dye wastewater using a microbial fuel cell coupled constructed wetland operating under different operating conditions. *Biosensors and Bioelectronics*, **68:** 135–141.

Galanopoulos, C., Sazakli, E., Leotsinidis, M. and Lyberatos, G. (2013). A pilot-scale study for modeling a free water surface constructed wetlands wastewater treatment system. *Journal of Environmental Chemical Engineering*, **1(4):** 642–651.

Gandy, C.J., Davis, J.E., Orme, P.H., Potter, H.A. and Jarvis, A.P. (2016). Metal removal mechanisms in a short hydraulic residence time subsurface flow compost wetland for mine drainage treatment. *Ecological Engineering*, **97:** 179–185.

Gazea, B., Adam, K. and Kontopoulos, A. (1996). A review of passive systems for the treatment of acid mine drainage. *Minerals Engineering*, **9(1):** 23–42.

Ghafari, S., Aziz, H.A., Isa, M.H. and Zinatizadeh, A.A. (2009). Application of response surface methodology (RSM) to optimize coagulation–flocculation treatment of leachate using polyaluminum chloride (PAC) and alum. *Journal of Hazardous Materials*, **163(2–3):** 650–656.

Gill, L.W., Ring, P., Casey, B., Higgins, N.M. and Johnston, P.M. (2017). Long term heavy metal removal by a constructed wetland treating rainfall runoff from a motorway. *Science of the Total Environment*, **601:** 32–44.

Glibert, P.M. (2017). Eutrophication, harmful algae and biodiversity—Challenging paradigms in a world of complex nutrient changes. *Marine Pollution Bulletin*, **124(2):** 591–606.

Gopal, B. (1999). Natural and constructed wetlands for wastewater treatment: Potentials and problems. *Water Science and Technology*, **40(3):** 27–35.

Griffiths, L.N. and Mitsch, W.J. (2017). Removal of nutrients from urban stormwater runoff by storm-pulsed and seasonally pulsed created wetlands in the subtropics. *Ecological Engineering*, **108:** 414–424.

Guo, L., Lv, T., He, K., Wu, S., Dong, X. and Dong, R. (2017). Removal of organic matter, nitrogen and faecal indicators from diluted anaerobically digested slurry using tidal flow constructed wetlands. *Environmental Science and Pollution Research*, **24(6):** 5486–5496.

Hanjra, M.A. and Qureshi, M.E. (2010). Global water crisis and future food security in an era of climate change. *Food Policy*, **35(5):** 365–377.

Hernández-Crespo, C., Gargallo, S., Benedito-Durá, V., Nácher-Rodríguez, B., Rodrigo-Alacreu, M.A. and Martín, M. (2017). Performance of surface and subsurface flow constructed wetlands treating eutrophic waters. *Science of the Total Environment*, **595:** 584–593.

Hijosa-Valsero, M., Matamoros, V., Sidrach-Cardona, R., Pedescoll, A., Martín-Villacorta, J., García, J. and Bécares, E. (2011). Influence of design, physico-chemical and environmental parameters on pharmaceuticals and fragrances removal by constructed wetlands. *Water Science and Technology*, **63(11):** 2527–2534.

Hijosa-Valsero, M., Matamoros, V., Martín-Villacorta, J., Bécares, E. and Bayona, J.M. (2010a). Assessment of full-scale natural systems for the removal of PPCPs from wastewater in small communities. *Water Research*, **44(5):** 1429–1439.

Hijosa-Valsero, M., Matamoros, V., Sidrach-Cardona, R., Martín-Villacorta, J., Bécares, E. and Bayona, J.M. (2010b). Comprehensive assessment of the design configuration of constructed wetlands for the removal of pharmaceuticals and personal care products from urban wastewaters. *Water Research*, **44(12):** 3669–3678.

Hoffmann, H., Platzer, C., Winker, M. and Muench, E. Von. (2011). Technology review of constructed wetland, subsurface flow constructed wetland for grey-water and domestic wastewater treatment. Deutsche Gesellschaft für Internationale Zusammenarbeit (GIZ) GmbH, Eschborn, Germany, 11.

Huang, J., Xu, C.C., Ridoutt, B.G., Wang, X.C. and Ren, P.A. (2017). Nitrogen and phosphorus losses and eutrophication potential associated with fertilizer application to cropland in China. *Journal of Cleaner Production*, **159:** 171–179.

Idris, A., Abdullah, A.G.L., Hung, Y.T. and Wang, L.K. (2014). Wetlands for wastewater treatment and water reuse. *In:* Modern Water Resources Engineering. Humana Press.

Jasper, J.T. and Sedlak, D.L. (2013). Phototransformation of wastewater-derived trace organic contaminants in open-water unit process treatment wetlands. *Environmental Science & Technology*, **47(19):** 10781–10790.

Jiang, Y., Martinez-Guerra, E., GnaneswarGude, V., Magbanua, B., Truax, D.D. and Martin, J.L. (2016). Wetlands for Wastewater Treatment. *Water Environment Research*, **88(10):** 1160–1191.

Jury, W.A. and Vaux, H.J. (2007). The emerging global water crisis: Managing scarcity and conflict between water users. *Advances in Agronomy*, **95:** 1–76.

Kadlec, R.H. (2009). Comparison of free water and horizontal subsurface treatment wetlands. *Ecological Engineering*, **35:** 159–74.

Kadlec, R.H. and Knight, R.L. (1996). Treatment Wetlands. New York: Lewis Publishers, CRC.

Kadlec, R.H. and Wallace, S.D. (2009). Treatment wetlands. 2nd Edition. CRC Press.

Karathanasis, A.D., Potter, C.L. and Coyne, M.S. (2003). Vegetation effects on fecal bacteria, BOD, and suspended solid removal in constructed wetlands treating domestic wastewater. *Ecological Engineering*, **20(2):** 157–169.

Khan, F.A. and Ansari, A.A. (2005). Eutrophication: An ecological vision. *The Botanical Review*, **71(4):** 449–482.

Khanitchaidecha, W., Sumino, T. and Kazama, F. (2010). Influence of carbon source on biological nitrogen removal by immobilised bacteria. *Journal of Water Resource and Protection*, **2(6):** 527.

Kivaisi, A.K. (2001). The potential for constructed wetlands for wastewater treatment and reuse in developing countries: A review. *Ecological Engineering*, **16(4):** 545–560.

Knight, R.L. (1997). Wildlife habitat and public use benefits of treatment wetlands. *Water Science and Technology*, **35(5):** 35–43.

Konnerup, D., Koottate, T. and Brix, H. (2009). Treatment of domestic wastewater in tropical, subsurface flow constructed wetlands planted with Canna and Heliconia. *Ecological Engineering*, **35(2):** 248–257.

Korkusuz, E.A., Beklioğlu, M. and Demirer, G.N. (2004). Treatment efficiencies of the vertical flow pilot-scale constructed wetlands for domestic wastewater treatment. *Turkish Journal of Engineering and Environmental Sciences*, **28(5):** 333–344.

Koskiaho, J., Ekholm, P., Räty, M., Riihimäki, J. and Puustinen, M. (2003). Retaining agricultural nutrients in constructed wetlands—experiences under boreal conditions. *Ecological Engineering*, **20(1):** 89–103.

Lee, C.G., Fletcher, T.D. and Sun, G. (2009). Nitrogen removal in constructed wetland systems. *Engineering in Life Sciences*, **9(1):** 11–22.

Lee, S.S., Paspalof, A.M., Snow, D.D., Richmond, E.K., Rosi-Marshall, E.J. and Kelly, J.J. (2016). Occurrence and Potential Biological Effects of Amphetamine on Stream Communities. *Environmental Science & Technology*, **50(17):** 9727–9735.

Liu, J., Liu, W., Wang, F., Kerr, P. and Wu, Y. (2016). Redox zones stratification and the microbial community characteristics in a periphyton bioreactor. *Bioresource Technology*, **204:** 114–121.

Liu, R., Zhao, Y., Zhao, J., Xu, L. and Sibille, C. (2017). Embedding constructed wetland in sequencing batch reactor for enhancing nutrients removal: A comparative evaluation. *Journal of Environmental Management*, **192:** 302–308.

Lu, S., Pei, L. and Bai, X. (2015). Study on method of domestic wastewater treatment through new-type multi-layer artificial wetland. *International Journal of Hydrogen Energy*, **40(34):** 11207–11214.

Madera-Parra, C.A., Peña-Salamanca, E.J., Peña, M.R., Rousseau, D.P.L. and Lens, P.N.L. (2015). Phytoremediation of landfill leachate with Colocasiaesculenta, Gynerumsagittatum and Heliconiapsittacorum in constructed wetlands. *International Journal of Phytoremediation*, **17 (1):** 16–24.

Matamoros, V. and Bayona, J.M. (2006). Elimination of pharmaceuticals and personal care products in subsurface flow constructed wetlands. *Environmental Science and Technology*, **40 (18):** 5811–5816.

Matamoros, V. and Salvadó, V. (2012). Evaluation of the seasonal performance of a water reclamation pond-constructed wetland system for removing emerging contaminants. *Chemosphere*, **86(2):** 111–117.

Matamoros, V., Arias, C., Brix, H. and Bayona, J.M. (2007). Removal of pharmaceuticals and personal care products (PPCPs) from urban wastewater in a pilot vertical flow constructed wetland and a sand filter. *Environmental Science & Technology*, **41(23):** 8171–8177.

Matamoros, V., Rodríguez, Y. and Bayona, J.M. (2017). Mitigation of emerging contaminants by full-scale horizontal flow constructed wetlands fed with secondary treated wastewater. *Ecological Engineering*, **99:** 222–227.

Matamoros, V., Sala, L. and Salvadó, V. (2012). Evaluation of a biologically-based filtration water reclamation plant for removing emerging contaminants: A pilot plant study. *Bioresource Technology*, **104:** 243–249.

Metcalf and Eddy (1991). Wastewater Engineering: Treatment Disposal Reuse (Third Edition). Mcgraw-Hill, Singapore.

Mitsch, W.J. and Wise, K.M. (1998). Water quality, fate of metals, and predictive model validation of a constructed wetland treating acid mine drainage. *Water Research*, **32(6):** 1888–1900.

Mojiri, A., Ziyang, L., Tajuddin, R.M., Farraji, H. and Alifar, N. (2016). Co-treatment of landfill leachate and municipal wastewater using the ZELIAC/zeolite constructed wetland system. *Journal of Environmental Management*, **166:** 124–130.

Morel, A. (2006). Greywater management in low and middle-income countries. Dubenforf, CH: Swiss Federal Institute of Aquatic Science and Technology.

Murray, K.E., Thomas, S.M. and Bodour, A.A. (2010). Prioritizing research for trace pollutants and emerging contaminants in the freshwater environment. *Environmental Pollution*, **158(12):** 3462–3471.

Nakamura, K., Hatakeyama, R., Tanaka, N., Takisawa, K., Tada, C. and Nakano, K. (2017). A novel design for a compact constructed wetland introducing multi-filtration layers coupled with subsurface superficial space. *Ecological Engineering*, **100:** 99–106.

Ngatia, L.W., Hsieh, Y.P., Nemours, D., Fu, R. and Taylor, R.W. (2017). Potential phosphorus eutrophication mitigation strategy: Biochar carbon composition, thermal stability and pH influence phosphorus sorption. *Chemosphere*, **180:** 201–211.

Nivala, J., Hoos, M.B., Cross, C., Wallace, S. and Parkin, G. (2007). Treatment of landfill leachate using an aerated, horizontal subsurface-flow constructed wetland. *Science of the Total Environment*, **380(1–3):** 19–27.

Ojoawo, S.O., Udayakumar, G. and Naik, P. (2015). Phytoremediation of phosphorus and nitrogen with Canna x generalis reeds in domestic wastewater through NMAMIT constructed wetland. *Aquatic Procedia*, **4:** 349–356.

Othman, E., Yusoff, M.S., Aziz, H.A., Adlan, M.N., Bashir, M.J. and Hung, Y.T. (2010). The effectiveness of silica sand in semi-aerobic stabilized landfill leachate treatment. *Water*, **2(4):** 904–915.

Pruden, A., Pei, R., Storteboom, H. and Carlson, K.H. (2006). Antibiotic resistance genes as emerging contaminants: Studies in northern Colorado. *Environmental Science & Technology*, **40(23):** 7445–7450.

Reddy, K.R., D'angelo, E.M. and DeBusk, T.A. (1990). Oxygen transport through aquatic macrophytes: The role in wastewater treatment. *Journal of Environmental Quality*, **19(2):** 261–267.

Renou, S., Givaudan, J.G., Poulain, S., Dirassouyan, F. and Moulin, P. (2008). Landfill leachate treatment: Review and opportunity. *Journal of Hazardous Materials*, **150(3):** 468–493.

Schmitt, N., Wanko, A., Laurent, J., Bois, P., Molle, P. and Mosé, R. (2015). Constructed wetlands treating stormwater from separate sewer networks in a residential Strasbourg urban catchment area: Micropollutant removal and fate. *Journal of Environmental Chemical Engineering*, **3(4):** 2816–2824.

Scholes, L., Shutes, R.B.E., Revitt, D.M., Forshaw, M. and Purchase, D.T. (1998). The treatment of metals in urban runoff by constructed wetlands. *Science of the Total Environment*, **214(1):** 211–219.

Sgroi, M., Pelissari, C., Ávila, C., Sezerino, P.H., Vagliasindi, F.G.A., García, J. and Roccaro, P. (2017). Removal of Conventional Water Quality Parameters, Emerging Contaminants and Fluorescing Organic Matter in a Hybrid Constructed Wetland System. *In:* Frontiers International Conference on Wastewater Treatment and Modelling, 2017.

Shelef, O., Gross, A. and Rachmilevitch, S. (2013). Role of plants in a constructed wetland: Current and new perspectives. *Water*, **5(2):** 405–419.

Shrestha, R.R., Haberl, R., Laber, J., Manandhar, R. and Mader, J. (2001). Application of constructed wetlands for wastewater treatment in Nepal. *Water Science and Technology*, **44 (11–12):** 381–386.

Song, Z.W., Wu, L., Yang, G., Xu, M. and Wen, S.P. (2008). Indicator microorganisms and pathogens removal function performed by copepods in constructed wetlands. *Bulletin of Environmental Contamination and Toxicology*, **81(5):** 459–463.

Stottmeister, U., Wießner, A., Kuschk, P., Kappelmeyer, U., Kästner, M., Bederski, O. and Moormann, H. (2003). Effects of plants and microorganisms in constructed wetlands for wastewater treatment. *Biotechnology Advances*, **22(1):** 93–117.

Tilley, E., Ulrich, L, Luethi, C., Reyomnd, P and Zurbruebb, C. (2014). Compendium of sanitation system and technologies. 2nd revised Edition. Switzerland: Swiss Federal Institute of Aquatic Science and Technology.

Tu, Y.T., Chiang, P.C., Yang, J., Chen, S.H. and Kao, C.M. (2014). Application of a CW system for polluted stream remediation. *Journal of Hydrology*, **510:** 70–78.

U.S. EPA (1999). Manual, constructed wetlands. Treatment of Municipal Wastewater. Washington D.C United States, EPA. url:https://nepis.epa.gov/Exe/ZyPDF.cgi/30004TBD.PDF?Dockey=30004TBD.PDF

Ulloa, M.J., Álvarez-Torres, P., Horak-Romo, K.P. and Ortega-Izaguirre, R. (2017). Harmful algal blooms and eutrophication along the Mexican coast of the Gulf of Mexico large marine ecosystem. *Environmental Development*, **22:** 120–128.

United Nations (2014). Department of Economic and Social Affairs, Publication Division. World Urbanization Prospects: The 2014 Revision. url:https://esa.un.org/unpd/wup/publications/files/wup2014-highlights.pdf

United States Environmental Protection Agency (2000). Guiding Principles for Constructed Treatment Wetlands: Providing for Water Quality and Wildlife Habitat. url:https://nepis.epa.gov/Exe/ZyPDF.cgi/2000536S.PDF?Dockey=2000536S.PDF

Villar, M.P., Domínguez, E.R., Tack, F., Ruiz, J.H., Morales, R.S. and Arteaga, L.E. (2012). Vertical subsurface wetlands for wastewater purification. *Procedia Engineering*, **42:** 1960–1968.

Vymazal J. (1995). Algae and Element Cycling in Wetlands. Chelsea: Lewis Publishers; p. 698.

Vymazal, J. (2001). Constructed wetlands for wastewater treatment in the Czech Republic. *Water Science and Technology*, **44(11–12):** 369–374.

Vymazal, J. (2005). Horizontal sub-surface flow and hybrid constructed wetlands systems for wastewater treatment. *Ecological Engineering*, **25(5):** 478–490.

Vymazal, J. (2007). Removal of nutrients in various types of constructed wetlands. *Sci Tot Env*, **380:** 48–65.

Vymazal, J. (2013). The use of hybrid constructed wetlands for wastewater treatment with special attention to nitrogen removal: A review of a recent development. *Water Research*, **47(14):** 4795–4811.

Vymazal, J. and Kröpfelová, L. (2011). A three-stage experimental constructed wetland for treatment of domestic sewage: First two years of operation. *Ecological Engineering*, **37(1):** 90–98.

Vymazal, J., Brix, H., Cooper, P., Green, M. and Haberl, R. (1998). Constructed wetlands for wastewater treatment in Europe: Backhuys Leiden. *Water Environ. Res*, **67(5):** 855–862.

Vymazal, J. (2010a). Constructed wetlands for wastewater treatment. *Water*, **2(3):** 530–549.

Vymazal, J. (2010b). Constructed wetlands for wastewater treatment: Five decades of experience. *Environmental Science & Technology*, **45(1):** 61–69.

Vymazal, J. (2002). The use of sub-surface constructed wetlands for wastewater treatment in the Czech Republic: 10 years experience. *Ecological Engineering*, **18(5):** 633–646.

Wallace, D.S. (2010). Application of constructed wetland for industrial wastewater treatment. Naturally Wallace Consulting. url= http://www.naturallywallace.com/docs/53_Application%20Of%20Industrial%20WW%20Treatment.pdf

Wang, M., Zhang, D.Q., Dong, J.W. and Tan, S.K. (2017). Constructed wetlands for wastewater treatment in cold climate—A review. *Journal of Environmental Sciences*, **57:** 293–311.

Wang, Q., Xie, H., Zhang, J., Liang, S., Ngo, H.H., Guo, W. and Li, H. (2015). Effect of plant harvesting on the performance of constructed wetlands during winter: Radial oxygen loss and microbial characteristics. *Environmental Science and Pollution Research*, **22(10):** 7476–7484.

Wang, X., Han, B.P., Shi, Y.Z. and Pang, Z.Q. (2009). Advanced wastewater treatment by integrated vertical flow constructed wetland with Vetiveria zizanioides in north China. *Procedia Earth and Planetary Science*, **1(1):** 1258–1262.

Weber, K.P. and Legge, R.L. (2008). Pathogen removal in constructed wetlands. *In:* Wetlands: Ecology, Conservation and Restoration, 176–211. Nova Science Publisher, Inc.

Wen, L.I.U., Hua, C.L., Ping, Z.Y. and Xiang, L.Z. (2011). Removal of total phosphorus from septic tank effluent by the hybrid constructed wetland system. *Procedia Environmental Sciences*, **10:** 2102–2107.

WWAP (United Nations World Water Assessment Programme) (2017). The United Nations World Water Development Report 2017, Wastewater, the Untapped Resource. Paris, UNESCO.

Yadav, A.K., Dash, P., Mohanty, A., Abbassi, R. and Mishra, B.K. (2012). Performance assessment of innovative constructed wetland-microbial fuel cell for electricity production and dye removal. *Ecological Engineering*, **47:** 126–131.

Yalcuk, A. and Ugurlu, A. (2009). Comparison of horizontal and vertical constructed wetland systems for landfill leachate treatment. *Bioresource Technology*, **100(9):** 2521–2526.

Yi, X., Tran, N.H., Yin, T., He, Y. and Gin, K.Y.H. (2017). Removal of selected PPCPs, EDCs, and antibiotic resistance genes in landfill leachate by a full-scale constructed wetlands system. *Water Research*, **121:** 46–60.

Zaharia, C. (2017). Decentralized wastewater treatment systems: Efficiency and its estimated impact against on-site natural water pollution status—A Romanian case study. *Process Safety and Environmental Protection*, **108:** 74–88.

Zhang, B.Y., Zheng, J.S. and Sharp, R.G. (2010). Phytoremediation in engineered wetlands: Mechanisms and applications. *Procedia Environmental Sciences*, **2:** 1315–1325.

Zhang, D.Q., Gersberg, R.M., Hua, T., Zhu, J., Goyal, M.K., Ng, W.J. and Tan, S.K. (2013). Fate of pharmaceutical compounds in hydroponic mesocosms planted with Scirpusvalidus. *Environmental Pollution*, **181:** 98–106.

Zhang, D.Q., Gersberg, R.M., Hua, T., Zhu, J., Tuan, N.A. and Tan, S.K. (2012). Pharmaceutical removal in tropical subsurface flow constructed wetlands at varying hydraulic loading rates. *Chemosphere*, **87(3):** 273–277.

Zhang, Q. and Weiming, L. (2011). Analysis on affected factors of treatment efficiency of rural sewage removal with constructed wetland. *Procedia Environmental Sciences*, **10:** 2314–2319.

Zhang, Z., Rengel, Z. and Meney, K. (2007). Nutrient removal from simulated wastewater using Canna indica and Schoenoplectus validus in mono- and mixed-culture in wetland microcosms. *Water, Air, and Soil pollution*, **183(1–4):** 95–105.

Chapter 3
Occurrence of Natural Radioactivity, Its Elevated Levels in Ground Water and Implications

Wedad Ali Abdurabu

1 Introduction

Ground water is the water that fills the natural open spaces in soil and rock underground. It is stored in the underground geological water system called an aquifer. An aquifer is any geological material that is filled with water and yields useful quantities of ground water to a well or spring. Both consolidated and unconsolidated geological materials are important as aquifers. Sedimentary rocks are the most important consolidated materials (bedrock), because they tend to have the highest porosities and permeability. Although most bedrock aquifers are within sedimentary rock, in some areas igneous or metamorphic rock can be important as aquifers (David et al. 1997). Sand and gravel aquifers are unconsolidated materials.

Most natural radionuclides are from U and Th series (Figs. 3.1 and 3.2). They are isotopes of eight different elements from thallium (Tl) up to U. Fortunately, in ground water few of them are soluble, mainly U, Ra and Rn. These radionuclides in groundwater are produced from rock-water interactions. There are many factors controlling the concentrations of these radionuclides in ground water, including the concentrations of the radioelements in the fractures of bedrocks that ground water move through, chemical and physical conditions of ground water that affect the water-rock interaction (Tricca et al. 2001; Reynolds et al. 2003).

Geology is the main factor controlling the level of radioactivity in the ground water. Deposits of U and Th were found in many rocks in trace amounts. Leaching of radioactive isotopes of ^{238}U, and ^{232}Th from these deposits into ground water occur due to water-rock interaction between these rocks and ground water (Bonotto and Bueno 2008).

W. A. Abdurabu (✉)
Department of Physics, Faculty of Science, Universiti Teknologi Malaysia, Skudai, Johore Bahru, Johore, Malaysia

National Atomic Energy Commission (NATEC), Sana'a, Yemen

D. Sengupta et al. (eds.), *Treatment and Disposal of Solid and Hazardous Wastes*,
https://doi.org/10.1007/978-3-030-29643-8_3

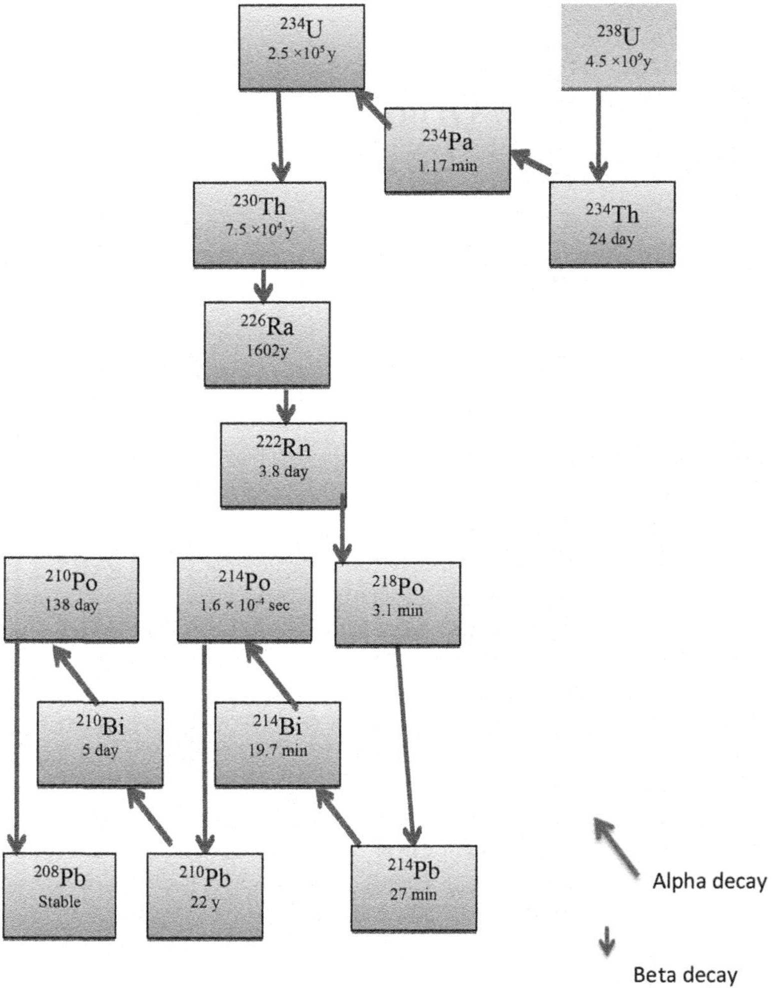

Fig. 3.1 Radioactive decay in ^{238}U series

High concentrations of natural radionuclides are related to elevated abundance of the parent radioelement in the associated rocks. For example, phosphate rocks usually have enhanced level of U, and the associated ground water has high concentrations of ^{226}Ra, a daughter of ^{238}U series. Wells constructed in granite bedrock have high concentration of ^{226}Ra and ^{228}Ra because granite rocks are enhanced with their parents U and Th (Vengosh et al. 2009).

Alpha particle emission is a physical factor affecting transferring of radionuclides into ground water, and it is called alpha recoil process. In this process, the ejection of an alpha particle from a radioactive nucleus causes the new nucleus to recoil up to several hundred nanometers in the opposite direction, depending on its surroundings.

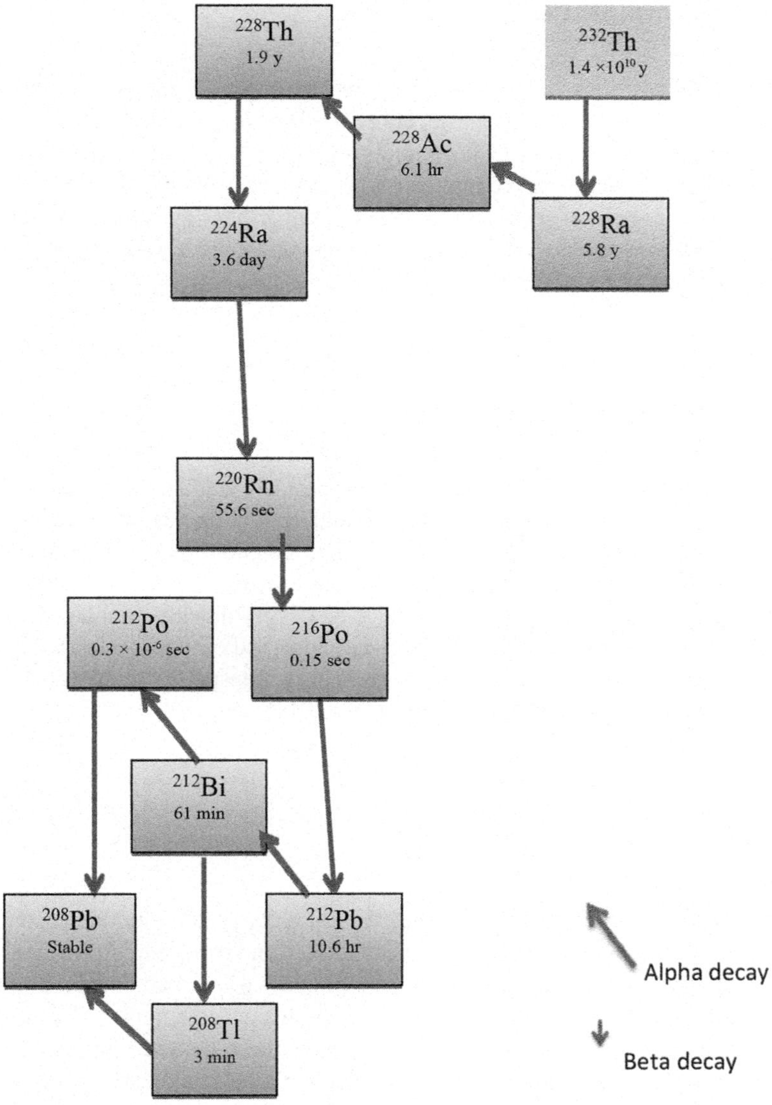

Fig. 3.2 Radioactive decay in ^{232}Th series

If the atom is near a phase boundary, the recoil can transport the atom across the boundary into a new environment, be it another grain or a pore filled with gas or liquid (Vengosh et al. 2009; Kraemer and Genereux 1998). The behaviour of most common natural radionuclides in ground water is presented. In addition, the health effects of radionuclides in ground water will be discussed and the international regulation of the concentration of radionuclides in drinking water will be presented.

1.1 K in Ground Water

Potassium readily dissolves in water by weathering and erosion of potassium-bearing minerals, such as feldspar. Its concentrations in ground water can vary in a wide range, depending on specific conditions in the aquifer and on the K content in the host rocks. The active concentration of ^{40}K can be calculated, taking into account its abundance in natural potassium and K content in the investigated water. High concentration levels of K were found to be corresponding to low pH values (Seghour and Seghour 2009; Chau et al. 2011).

1.2 Th in Ground Water

One of the most insoluble radioactive elements in water is Th. It has only one oxidation state in nature (+4). Th forms insoluble hydroxides under most natural conditions because of its high charge and ionic potential (Mason 1966). For this reason, Th is rare in natural waters and this element is not transported very far in solution, but it is soluble in strongly acidic water. Very quickly after it is weathered out of a rock or other solid phase, Th becomes adsorbed onto colloidal or clay-sized particles and remains near its original location, or is transported by physical erosion as suspended sediment in streams (Chau et al. 2011).

1.3 U in Ground Water

Compared to Th, U is soluble in water. U has a fairly complex chemistry, with correspondingly complex behaviour in near-surface and surface hydrologic systems. U can be present in the oxidation states +6, +5, and + 4. In oxidizing environments, the +6 state is predominant and forms soluble complexes with hydroxide, carbonate, fluoride, sulphate or phosphate, depending upon the type and amount of anions present in solution and pH of the solution (Almeida et al. 2004; Kraemer and Genereux 1998). For example, the solubility of U increased by formation of uranyl carbonate complexes above pH 8.0 and below pH 5 by formation of uranyl phosphate, or uranyl fluoride (Durrance 1986).

Under progressively reducing conditions, U is reduced to +5 and + 4. This generally occurs as free oxygen is removed from the system, followed by the appearance of hydrogen sulfide. The result is the removal of U from solution due to the precipitation of insoluble U minerals. The chemical characteristics described above allow U to be weathered from rocks, transported long distances in solution, and then deposited by various means far from the site of weathering, or remain in solution for long periods of time (Kraemer and Genereux 1998).

1.4 Ra in Ground Water

There are two Ra isotopes of concern in drinking water: ^{226}Ra is the fifth member of ^{238}U series and decays by alpha emission with half-life of 1622 years; ^{228}Ra is the second member of ^{232}Th and decays by beta emission with half-life of 5.7 years. Several processes are responsible for Ra input to the groundwater processes including decay of dissolved parent isotopes, alpha-recoil, desorption from aquifer surfaces, dissolution of aquifer solids, ion exchange, and leaching from radiation-damaged crystals (Sturchio et al. 2001; Kiro et al. 2015).

Radium gets into groundwater due to many different processes such as dissolution of Ra from the aquifer's martials, alpha recoil due to decay of its parent Th in host rock, desorption from the aquifer surface, and decay of Th in the solution (Fig. 3.3) (Kiro et al. 2015; Porcelli 2008). It gets out the water by several processes including radioactive decay, adsorption and precipitation. Radium could be considered intermediate in reactivity between Th and U in surface and near-surface environments. It has only one oxidation state (+2) and forms only weak complexes in solution. It is, however, very insoluble as a sulfate and will precipitate with Ba or Ca (Langmuir and Riese 1985). In most aqueous situations, Ra is subject to adsorption onto silicates, clays, and oxyhydroxides, but not as strongly as Th. It is therefore detectable in most water (Kraemer and Genereux 1998). In groundwater with low degree of salinity, the exchange reaction with clay minerals affects the concentrations of Ra (Vengosh et al. 2009; Szabo et al. 2012).

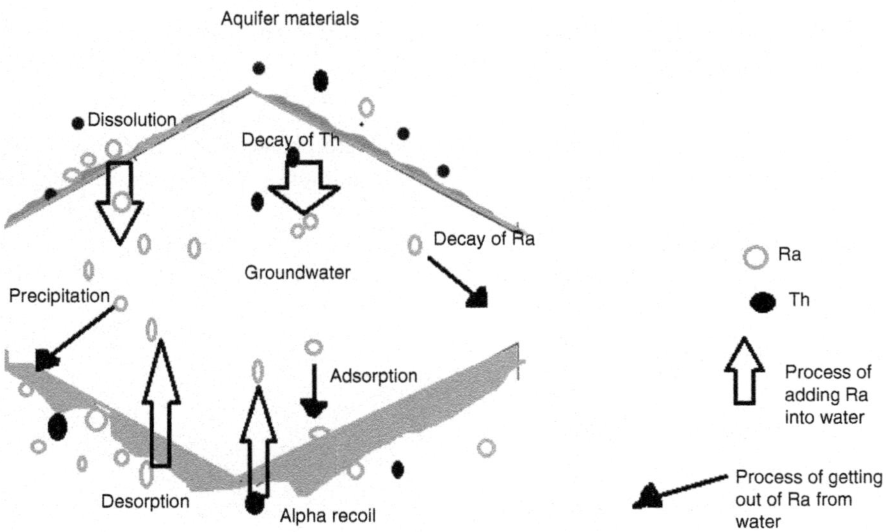

Fig. 3.3 Geochemical behaviour of Ra in groundwater

1.5 Rn in Ground Water

There are three natural isotopes of radon (^{219}Rn, ^{220}Rn and ^{222}Rn), but only ^{222}Rn is of importance in groundwater systems. It is a decay product of ^{226}Ra, ^{222}Rn decays to ^{218}Po with a half-life of 3.823 days. Rn concentration levels are very low in surface water, while the highest concentration of Rn is usually found in groundwater. Rn concentration in groundwater is mainly controlled by the lithology of the aquifer. Rn is a noble gas and begins to diffuse through the crystalline lattice of the enclosing mineral immediately after being formed from the decay of radium. Rn can therefore easily reach small cracks and imperfections in the crystal and migrates into larger pore spaces filled with water and gas. The process by which Rn escapes from solid material is called emanation. The emanating power of a solid is the fraction of Rn atoms formed within it that escapes to the outside. Small grain-size sediments usually have a greater emanating power than coarse-grain sediments, as the surface-area to the volume ratio is greater. Similarly, highly-fractured granites and metamorphic rocks have higher emanating power than sparsely fractured granite. The fact that granites are often fractured (high emanating power) coupled with the high U content of granite (high source of Rn) means water from granitic terrain often has very high Rn concentrations. High concentrations of ^{222}Rn could lead to detectable levels of ^{210}Pb and ^{210}Po (Kraemer and Genereux 1998; Chau et al. 2011; Isam et al. 2002; Bonotto and Caprioglio 2002; Ali et al. 2010).

2 Physicochemical Parameters of Water Controlling Radionuclides Levels in Water

The physical and chemical properties of water are characteristic of the climatic, geochemical, geomorphological and pollution conditions prevailing in the underlying aquifer. The chemical condition of the water controls the mobilization of radionuclides in groundwater from aquifer rocks. The level of acidity (pH), salinity and temperature of groundwater control the leaching of natural radionuclides into the groundwater.

2.1 Temperature Effect (T)

The rate of many biological and chemical reactions is affected by temperature, though deep groundwater temperatures are less susceptible to seasonal temperature fluctuations. Temperature should be recorded to 0.1 °C (Trick et al. 2018). High temperatures of groundwater affect the water-rock interaction, which lead to leaching of minerals from aquifer's rock into groundwater. Hot springs are characteristically enriched in Rn and Ra (Durrance 1986). In addition, it was found in the geopressured aquifers of the U.S. Gulf Coast that Ra concentration in groundwater increased with high temperature (Kraemer and Genereux 1998). In sandstone aquifer in elevated area in Juban, Yemen, ^{228}Ra correlates moderately with water temperature because of reequilibration of silica between solid phase and water (Abdurabu et al. 2016b). At high temperature and salinity of water, the exchange rate of silica between solid phase and water increase, which uncover Ra atoms in solid phase allowing for easy entry into water (Kraemer 1991).

2.2 *pH Effect*

pH is a measure of the hydrogen ion concentration in solution and is also referred to as the degree of acidity or alkalinity. As a sample's pH changes, many precipitation, co-precipitation and sorption processes can occur that alter the sample's chemical composition and reaction rates (Trick et al. 2018). pH is considered as an important ecological factor and provides important information on many types of geochemical equilibrium or solubility calculation (Prasad et al. 2014). The pH of water plays an important role on the occurrence of radionuclide in water.

The radionuclides concentration in lower pH waters are more than in high pH waters (Osmond and Cowart 2000). In high background radiation area in Saudi Arabia, the concentration of U was increased in the low pH waters (Shabana and Kinsara 2014). Th is strongly sorbed onto kaolinite and quartz, among other solids, at pH values greater than 4 (Kraemer and Genereux 1998; Langmuir and Herman 1980; Senior and Vogel 1995). In elevated area in Juban District, Yemen, ^{228}Ra has a moderate negative correlation with pH (Abdurabu et al. 2016b). This is in agreement with many studies of sandstone aquifers where high values of Ra are associated with low pH (Senior and Vogel 1995; Kumar et al. 2016).

2.3 *Total Dissolved Solids (TDS)*

Total dissolved solids (TDS) is a measure of all dissolved substances in water. TDS is measured in a laboratory and reported as mg l^{-1} (Salem and El-Sayed 2015). There is correlation between Ra concentration in groundwater and TDS. At TDS levels of greater than 1000 ppm, Ra solubility increased due to the common ion effect. In this process, the absorbed Ra atoms could be replaced by more common ions in water.

Uranium concentrations were found to increase sharply as the TDS of water increased in groundwater of Juban district, Yemen and Punjab state, India (Abdurabu et al. 2016b; Kumar et al. 2014). This indicated that at elevated TDS, the competition for sorption sites between ions in groundwater is high and adsorbed U atoms may be replaced by more common ions in water (Abdurabu et al. 2016b; Kumar et al. 2014).

2.3.1 Electrical Conductivity (EC)

Electrical conductivity (EC) is a measure of water capacity to convey electric current. It depends upon the presence of charged ion species such as calcium, sodium, potassium, magnesium chloride, etc. It signifies the amount of total dissolved salts and is a useful tool to evaluate the purity of water. Pure water has a low electrical conductivity, but this increases significantly with the dissolution of a small amount of ionic material such as sodium chloride. In general, the salinity of water increases as EC increase (Trick et al. 2018; Prasad et al. 2014).

The EC enhanced with increasing water-rock interaction, which affected the concentrations of radionuclides in groundwater. There was inverse relation between EC and Rn concentration in South Korea's groundwater (Cho et al. 2015). In the groundwater of Chickies Quartizite, Southwestern Pennsylvania, US, Ra concentrations were correlated positively with EC because ion solubility increased with high conductance and ionic strength (Senior and Vogel 1995).

Electrical conductivity is an indicator of the salinity of groundwater. There is a positive correlation between the degree of salinity and U concentrations. U desorption increased by increasing the water salinity (Abdurabu et al. 2016a). In groundwater of high background area in Juban, Yemen, there was very positive strong correlation between Rn and EC which indicated a strong relation between activity concentration of ^{222}Rn and the degree of salinity. The results were in agreement with Moise et al. (2000), who concluded that strong changes in the salinity of groundwater could cause ^{226}Ra to deposit on aquifer surfaces and be a local source of ^{222}Rn (Porcelli 2008; Abdurabu et al. 2016a; Moise et al. 2000).

3 Radiological Health Effects

Exposure to radionuclides leads to increased risk of cancer. The radioactive particles (alpha, beta) and gamma ray emitted by radionuclides are called "ionizing radiation" because they ionize nearby atoms as they travel through a cell or other material. In living tissue, this ionization process can damage chromosomes or other parts of the cell. This cellular damage can lead to the death of the cell or to unnatural reproduction of the cell. When a cell reproduces uncontrollably, it becomes a cancer (UNSCEAR 2000; WHO 2008).

The radiation dose resulting from ingestion of a radionuclide depends on a number of chemical and biological factors. These include the fraction of the intake that is absorbed from the gut, the organs or tissues to which the radionuclide is transported and the time during which the radionuclide remains in the organ or tissue before excretion. The nature of the radiation emitted on decay and the sensitivity of the irradiated organs or tissues to radiation must also be considered (WHO 2008; ICRP 1996; USEPA 1999).

In the case of water, activity concentration is given in Bq L^{-1}. This value can be related to an effective dose per year (mSv $year^{-1}$) using a dose coefficient (mSv Bq^{-1}) and the average annual consumption of water litres per year (l $year^{-1}$). The effective dose arising from the ingestion of a radioisotope in a particular chemical form can be estimated using a dose coefficient (WHO 2008; ICRP 1996; USEPA 1999).

The elements that cause radiation exposure in water are U, Ra and Rn. U nuclides emit alpha rays of high ionisation power and, therefore, it may be hazardous if inhaled or ingested in higher quantity. The EPA has established maximum contaminant levels (MCLs) of elements and radionuclides in drinking water by balancing the health hazards due to consumption specific element in drinking water compared to the availability of water source and the cost of reducing the concentrations of these elements in water sources (ICRP 1996; Ranil et al. 2013).

3.1 Health Standards and Criteria for U in Water

According to an estimate, food contributes ≈15% of ingested uranium, while drinking water contributes ≈85% (Hadad and Doulatdar 2008). Adverse health effects from natural U can be due to its radioactive and chemical properties. Radioactive effects are very small from natural U; chemically it can be harmful to the kidneys from large exposure (ICRP 1996; Hadad and Doulatdar 2008). The EPA set 30 μg L^{-1} as MCL for U in drinking water, which is based on its chemical toxicity to the kidney and damage it (Ranil et al. 2013).

3.2 Health Standards and Criteria for Ra in Water

Radium is of radiological importance because Ra behaves chemically like calcium. It accumulates in the same organs as do calcium and, thus, is a bone seeker (Wanty and Schoen 1991). It might be expected that Ra would induce leukemia through its radiation effects on bone marrow; however, this effect appears to be insignificant compared to the number of bone sarcomas Ra has caused (Bonotto and Caprioglio 2002). The EPA MCL for Ra is 185 mBq L^{-1} based on its potential to cause cancer (Ranil et al. 2013).

3.3 Health Standards and Criteria for Rn in Water

Exposure to Rn and its progeny is assumed to be associated with increased risks of several kinds of cancer (Ranil et al. 2013). When Rn or its progeny are inhaled, lung cancer accounts for most of the total incremental cancer risk, while ingestion of radon in water is suspected of being associated with increased risk of tumours of several internal organs, primarily the stomach (Ranil et al. 2013; Alabdula'aly 2014; USEPA 2007; Moreno et al. 2014; Abdurabu 2017).

The EPA has provisionally established that MCL for Rn is 11.1 Bq L^{-1} based on its effect on lungs that lead to cause lung cancer. The risk of Rn exposure is increased due to indoor Rn, the source of which is the migration of Rn in soil gas into houses, and the emission and consumption of Rn from drinking water. Consequently, the EPA has recommended an Alternative Maximum Contaminant Level (AMCL) of Rn in water. In case that Rn in indoor air was reduced, the AMCL of Rn in water is 148 Bq L^{-1} (ICRP 1996; (Ranil et al. 2013).

References

Alabdula'aly, A.I. (2014). Occurrence of radon in groundwater of Saudi Arabia. *Journal of Environmental Radioactivity*, **138:** 186–191. Elsevier.

Abdurabu, W.A., Ramli, A.T., Saleh, M.A. and Heryansyah, A. (2016a). The activity concentrations of ^{222}Rn and corresponding health risk in groundwater samples from basement and sandstone aquifer; the correlation to physicochemical parameters. *Radiation Physics and Chemistry*, **127:** 34–41. Elsevier.

Abdurabu, W.A., Ramli, A.T., Saleh, M.A. and Heryansyah, A. (2016b). Occurrence of natural radioactivity and corresponding health risk in groundwater of an elevated radiation background in Juban District, Yemen. *Journal of Environmental Earth*, **36:** 163–177.

Abdurabu, W.A. (2017). Natural Radioactivity levels of Selected Areas in Juban District Yemen. PhD Thesis. Universiti Teknologi Malaysia, Johor, Malaysia.

Ali, N., Akhter, P., Khan, F. and Waheed, A. (2010). Estimation of mean annual effective dose through radon concentration in the water and indoor air of Islamabad and Murree. *Radiation Protection Dosimetry*, **141(2):** 183–191. Oxford University Press.

Almeida. R.M.R., Lauria, D.C., Ferreira, A.C. and Sracek, O. (2004). Groundwater radon, radium and uranium concentrations in Região dos Lagos, Rio de Janeiro State, Brazil. *Journal of Environmental Radioactivity*, **73:** 323–334. Elsevier.

Bonotto, D.M. and Bueno, T.O. (2008). The natural radioactivity in Guarani aquifer groundwater, Brazil. *Applied Radiation and Isotopes*, **66(10):** 1507–1522. Elsevier.

Bonotto, D.M. and Caprioglio, L. (2002). Radon in groundwaters from Guarany aquifer, South America: Environmental and exploration implications. *Applied Radiation and Isotopes*, **57(6):** 931–940. Elsevier.

Chau, N.D., Dulinski, M., Jodlowski, P., Nowak, J., Rozanski, K., Sleziak, M. and Wachniew, P. (2011). Natural radioactivity in groundwater – a review. *Isotopes in Environmental and Health Studies*, **47(4):** 415–437. Taylor & Francis Publisher.

Cho, B.W., Choo, C.O., Kim, M.S. and Hwang, J. (2015). Spatial relationships between radon and topographical, geological, and geochemical factors and their relevance in all of South Korea. *Environ Earth Sci.*, **74:** 5155–5168.

David, R., Larson, J.P.K. and Scott, M. (1997). Geologic, Geophysical, and Hydrologic Investigations for a Supplemental Municipal Groundwater Supply, Danville, Illinois. Cooperative Groundwater Report 18, the State of Illinois.

Durrance, E.M. (1986). Radioactivity in Geology: Principles and Applications. Wiley & Sons, New York.

Hadad, K. and Doulatdar, R. (2008). U-series concentration in surface and ground water resource of Ardabil Province. *Radiation Protection Dosimetry*, **130(3):** 309–318. Oxford University Press.

International Commission on Radiological Protection (ICRP) (1996). Age-dependent doses to members of the public from intake of radionuclides: Part 5. Compilation of Ingestion and Inhalation Dose Coefficients. ICRP Publication 72.

Isam, S.M., Pettersson, H.B.L. and Lund, E. (2002). Uranium and thorium series radionuclides in drinking water from drilled bedrock wells, correlation to geology and bedrock radioactivity, and dose estimation. *Radiat. Prot. Dosim.* **102(3):** 249–258. Oxford University Press.

Kiro, Y., Weinstein, Y., Starinsky, A. and Yechieli, Y. (2015). Application of radon and radium isotopes to groundwater flow dynamics: An example from the Dead Sea. *Chemical Geology*, **411:** 155–171.

Kraemer, T.F. (1991). Uranium, Radium, and Radon in deeply buried sediments of the U.S. Gulf Coast. *In:* Gundersen, L.CS, Wanty, R.B (Eds). In Field Studies of Radon in Rocks, Soils, and Water, U.S. Geological Survey Bulletin 1971.

Kraemer, T.F. and Genereux, D.P. (1998). Applications of Uranium- and Thorium-Series Radionuclides in Catchment Hydrology Studies. *In:* Kendall, C. and McDonnell, J.J. (Eds). Isotope Tracers in Catchment Hydrology. Amsterdam: Elsevier.

Kumar, A., Tripathi, R.M., Rout, S., Mishra, M.K., Ravi, P.M. and Ghosh, A.K. (2014). Characterization of groundwater composition in Punjab state with special emphasis on uranium content, speciation and mobility. *Radiochim. Acta*, **102(3):** 239–254.

Kumar, A., Karpe, R.K., Rout, S., Gautam, Y.P., Mishra, M.K., Ravi, P.M. and Tripathi, R.M. (2016). Activity ratios of $^{234}U/^{238}U$ and $^{226}Ra/^{228}Ra$ for transport mechanisms of elevated uranium in alluvial aquifers of groundwater in southwestern (SW) Punjab, India. *Journal of Environmental Radioactivity*, **151:** 311–320.

Langmuir, D. and Herman, J. (1980). The mobility of thorium in natural waters at low temperatures. *Geochim. et Cosmochim. Acta.*, **44:** 1753–1766.

Langmuir, D. and Riese, A. (1985). The thermodynamic properties of radium. *Geochim. et Cosmochim. Acta.*, **49:** 1593–1601.

Mason, B. (1966). Principles of Geochemistry. New York, Wiley.

Moise, T., Starinsky, A., Katz, A. and Kolodny, Y. (2000). Ra isotopes and Rn in brines and ground waters of the Jordan-Dead Sea Rift Valley: Enrichment, retardation, and mixing. *Geochimica et Cosmochimica Acta.*, **64:** 2371–2388.

Moreno, V., Bach, J., Baixeras, C. and Font, L. (2014). Radon levels in groundwaters and natural radioactivity in soils of the volcanic region of La Garrotxa, Spain. *Journal of Environmental Radioactivity*, **128:** 1–8.

Osmond, J.K. and Cowart, J.B. (2000). U-series nuclides as tracers in groundwater hydrology. *In:* Cook, P.G. and Herczeg, A.L. (Eds). Environmental Tracers in Subsurface Hydrology, Springer.

Porcelli, D. (2008). Investigating Groundwater Processes Using U- and Th-Series Nuclides. *In:* Krishnaswami, S., Cochran, J.K. (Eds). Radioactivity in the Environment: U-Th Series Nuclides in Aquatic Systems. First ed, vol 13. Elsevier.

Prasad, M., Reddy, M.B., Reddy, M.R. and Sunitha,V. (2014). Studies on physicochemical parameters to assess the water quality in Obulavaripalli Mandal of YSR (Kadapa) District, Andhra Pradesh, India. *Int. J. Curr. Res. Aca. Rev.*, **2(12):** 31–41.

Ranil, A., Duggal, V. and Balaram, V. (2013). Uranium Estimation in Drinking water Samples from some Areas of Punjab and Himachal Pradesh, India Using ICP-MS. *Radiation Protection Dosimetry*, **157(1):** 146–151.

Reynolds, B.C., Wasserburg, G.J. and Baskaran, M. (2003). The transport of U- and Th-series nuclides in sandy confined aquifers. Geochimica et Cosmochimica Acta., **67(11):** 1955–1972.

Salem, W.M. and El-Sayed, M. (2015). Hydro-geochemical and isotopic composition of ground water in Helwan area. *Egyptian Journal of Petroleum*, **24:** 411–421.

Seghour, A. and Seghour, F.Z. (2009). Radium and ^{40}K in Algerian Bottled Mineral Waters and Consequent Doses. *Radiat. Prot. Dosim.*, **133(1):** 50–57.

Senior, L.A. and Vogel, K.L. (1995). Radium and Radon in Ground Water in the Chickies Quartzite, Southeastern Pennsylvania. U.S. Geological Survey Water-Resources Investigations Report 92-4088, Lemoyne, Pennsylvania.

Shabana, E.I. and Kinsara, A.A. (2014). Radioactivity in the groundwater of a high background radiation area. *Journal of Environmental Radioactivity*, **137:** 181–189.

Sturchio, N.C., Banner, J.L., Binz, C.M., Heraty, L.B. and Musgrove, M. (2001). Radium geochemistry of ground waters in Paleozoic carbonate aquifers, midcontinent, USA. *Applied Geochemistry*, **16:** 109–122.

Szabo, Z., DePaul, V.T., Fischer, J.M., Kraemer, T.F. and Jacobsen, E. (2012). Occurrence and geochemistry of radium in water from principal drinking-water aquifer systems of the United States. *Applied Geochemistry*, **27:** 729–752.

Tricca, A., Wasserburg, G.J., Porcelli, D. and Baskaran, M. (2001). Transport of U- and Th-series nuclides in a sandy unconfined aquifer. *Geochimica et Cosmochimica Acta.*, **65(8):** 1187–1210.

Trick, J.K., Stuart, M. and Reeder, S. (2018). Contaminated groundwater sampling and quality control of water analyses. *In:* Benedetto De Vivo, Harvey E. Belkin, Annamaria Lima (eds). *Environmental Geochemistry* (2nd Edition), 25–45. Elsevier. ISBN 9780444637635.

UNSCEAR (United Nations Scientific Committee on the Effects of Atomic Radiation, Sources) (2000). Effects and Risks of Ionizing Radiation. Report to the General Assembly with Annex B. United Nations, New York.

USEPA (1999). Radon in Drinking Water Health Risk Reduction and Cost Analysis. USEPA (United States Environmental Protection Agency), Washington, Federal Register, **64:** 9559–9599.

USEPA (2007). Drinking Water Standards and Health Advisories Table. http://www3.epa.gov/region9/water/drinking/files/dwsha_0607.pdf. .

Vengosh, A., Hirschfeld, D., Vinson, D., Dwyer, G., Raanan, H., Rimawi, O., Al-Zoubi, A., Akkawi, E., Marie, A., Haquin, G., Zaarur, S. and Ganor, J. (2009). High naturally occurring radioactivity in fossil groundwater from the Middle East. *Environmental Science and Technology*, **43:** 1769–1775.

Wanty, R.B. and Schoen, R. (1991). A review of the chemical processes affecting the mobility of radionuclides in natural waters, with applications. *In:* Gundersen, L.C.S. and Wanty, R.B. (Eds). Field Studies of Radon in Rocks, Soils, and Water. U.S. Geological Survey Bulletin 1971, 183–192.

World Health Organization (WHO) (2008). Guidelines for drinking-water quality [electronic resource]: incorporating 1st and 2nd addenda, Recommendations. World Health Organization.

Chapter 4
Refuse Derived Fuel (RDF) Production and Utilisation Potential from Municipal Solid Waste (MSW) in India

Dirk Weichgrebe, Christopher Speier, Moni Mohan Mondal, and Shivali Sugandh

1 Introduction

India had the second largest population in the world in 2017 with 1.34 billion inhabitants, and its population-specific CO_2 emissions rose from 0.98 in 2000 to 1.73 tons per capita in 2014 (The World Bank 2018). Indian urban areas generated 62 million tons of municipal solid waste (MSW) in 2011 (Planning Commission 2014). While 50 to 90 wt.-% of MSW are collected, in average almost 87 wt.-% is disposed on landfills without further processing or treatment (PIB 2016; CPCB 2016; Banerjee 2016; Mani and Singh 2016). A review of urban MSW in India shows that approx. 51 wt.-% (±12.4) are comprised of organic materials, whereas plastics and papers contributed in average 6.7 (±4.4) and 7.7 (±5.5) wt.-% respectively (Speier et al. 2018b).

For many urban local bodies (ULBs) in India, the collection, transport, treatment and disposal of MSW remains a challenge. Solid Waste Management (SWM) still causes adverse impacts on the environment and public health not at least from the identified 2230 open dumpsites in India (CPCB 2018). However, MSW in India also offers potential for material, energy and nutrient recovery. The aim of future SWM infrastructure development in India must therefore be to exploit this potential in order to improve public life and health, and the environment (Joshi and Ahmed 2016; Kumar et al. 2017).

The Indian Integrated Solid Waste Management (ISWM) hierarchy gives emphasis to prevention, reduction, reuse, recycling, recovery and disposal, with prevention being the most preferred option and the disposal at the landfill being the least. Thereby material recycling of non-biodegradable waste stream and composting of separately collected organic waste is preferable to energy recovery from waste

D. Weichgrebe (✉) · C. Speier · M. M. Mondal · S. Sugandh
Institute of Sanitary Engineering and Waste Management, Leibniz Universität Hannover, Hannover, Germany
e-mail: weichgrebe@isah.uni-hannover.de

D. Sengupta et al. (eds.), *Treatment and Disposal of Solid and Hazardous Wastes*,
https://doi.org/10.1007/978-3-030-29643-8_4

(material recovery is not possible) by incineration, pyrolysis, bio-methanisation or co-combustion of refuse derived fuel (RDF) in waste to energy plants (WtE), cement kilns or other industries. India has taken several policy initiatives to promote WtE technologies like incineration, RDF based power plants or co-incineration of RDF in various sectors (such as cement, coal, steel, etc.). For instance, grants and capital subsidies upto 30 million INR per megawatt is provided for installing waste to energy plants by Ministry of New and Renewable Energy (MNRE 2018). The SWM rules 2016 also enforce compulsory purchase of power generated from waste to energy plants at a preferential tariff of 7.04 INR for MSW and 7.90 INR for RDF power plants (CERC 2015). However, these political efforts have so far been less successful.

In order to avoid the disposal of the non-recyclable combustible fraction of MSW in landfills (e.g. contaminated paper, cloth and plastics, multilayer and packaging materials, leather, rubber, and tyre pieces), it may be utilized as an alternative fuel in the cement industry. This is in line with India's Intended Nationally Determined Contributions (INDC), where cement industry (2nd largest greenhouse gas (GHG) contributor after electricity) has a potential to reduce its carbon footprint by thermal substitution of 20% or more of conventional fuel by 2022 (CII 2016). Currently the cement industry in India is co-processing different types of waste including biomass, hazardous and non-hazardous waste with a Thermal Substitution Rate (TSR) of approx. 2.5% (CPHEEO 2018).

2 Definition of RDF

The early development of RDF technology occurred mainly in the UK and to some extent in Italy, with plants built from the mid-1970s onward (McDougall et al. 2003). However, although the term RDF is internationally widely used, definitions vary across the globe. Most definitions coincide that RDF is considered as the processed high calorific fraction, also labelled as combustible fraction of MSW, commercial and industrial waste (Kara et al. 2009; Rotter 2010; Rada and Andreottola 2012; Gallardo et al. 2014), but remain vague in terms of product quality parameters. RDF generally describes unspecified waste after basic processing to increase the calorific value of these waste categories (WRAP 2012).

The SWM Rules 2016 in India defines RDF as "fuel derived from combustible waste fraction of solid waste like plastic, wood, pulp or organic waste, other than chlorinated materials, in the form of pellets or fluff produced by drying, shredding, dehydrating and compacting of solid waste". This combustible waste fraction is defined as non-biodegradable, non-recyclable, non-reusable, non-hazardous solid waste having minimum calorific value exceeding 1500 kcal/kg (6.276 MJ/kg), excluding chlorinated materials like plastic, wood pulp, etc.

Other terminologies for waste derived fuels are secondary fuels, substitute fuels, solid recovered fuels (SRF) or alternative fuels (Velis et al. 2010). The International Finance Corporation (IFC 2017) summarizes common quality characteristics of

RDF with Net Calorific Value (NCV) of 8–15 MJ/kg, moisture content (MC) of 25–40 wt.-% and a common particle size (x_{50}) of 0–400 mm.

The RDF is characterised by a commonly heterogeneous mixture of various waste materials, derived from MSW, old tyres, old wood, sewage sludge etc. The segregated, combustible fraction of MSW comprises materials like packaging material, textiles, shoes, plastic bottles and containers, paper, cardboard, other types of plastics, rubber, composite and organic materials. Each material has individual characteristics regarding particle size, bulk density, elasticity, brittleness, heating value, combustion behaviour, physical and chemical properties. Besides NCV, MC and x_{50} key parameters for the classification of RDF are the contents of ash, chlorine and sulphur.

Within a clearly defined range, facilities utilising RDF are usually able to cope with fluctuations in RDF composition due to the heterogeneity of the materials. However, the broad variety of material mixtures labelled as RDF poses a challenge for the design of technical mono- or co-combustion plants. Moreover, thermal treatment of waste must comply with national air emission standards in most countries. Although the production of RDF may offer an attractive opportunity to both reduce the waste amount and substitute fossil fuels, certain quality parameters need to be ensured to meet legal, environmental, and technical requirements.

The chlorine content is therefore one of the main factors in the quality determination of substitute fuels, since chlorine causes technical problems such as high-temperature corrosion during combustion and causes increased expenditure for flue gas cleaning (Hase et al. 2014). In MSW based RDF, the highest concentrations of chlorine are found in non-packaging plastics (up to 6 wt.-% dry basis), textiles or shoes due to the presence of polyvinylchloride (PVC) (Ma et al. 2008). Another major source for high chlorine levels in RDF is found to be chloride salts of kitchen waste (Ma et al. 2010). The impact of high chlorine contents on the process depends on the bond type (organic or inorganic) and the interaction with alkalis, sulphur and heavy metals (Beckmann and Ncube 2007). Generally, low chlorine levels are preferred for the production of RDF (Hase et al. 2014). As stated by Hanjer Biotech HBEPL, waste pickers in India already remove large shares of PVC from MSW before it enters the RDF processing unit (GIZ-IGEP 2013). In EU and Germany PVC recycling is increasing, in particular PVC in building products (AGPU 2018). However, for an efficient utilisation of RDF and guaranteed product qualities, chlorinated materials have to be removed during processing.

The same applies to metal-containing materials such as cans, batteries or cables, some of which contain high concentrations of heavy metals (Skutan and Brunner 2012). With increasing concentrations of heavy metals, flue gas cleaning requires greater technical and financial effort, which affects the viability of RDF utilization. In cement kilns, most heavy metals with the exception of the volatile elements mercury and thallium are bound in the clinker during the combustion process (Velis et al. 2010), but in combustion processes these remain in the slag. Plant operators require guaranteed RDF qualities in order to maintain a technically smooth and economically viable operation of the incineration process. The targeted processing of RDF and an unequivocal monitoring and quality assurance system therefore is of similar importance for the sustainable and environmental sound utilization of RDF in Indian waste management.

3 Legislation and Standards on RDF

3.1 Legislation and Standards in India

The legal and regulatory framework for waste management in India began in late 1980s when the first Hazardous Waste (Management & Handling) Rules 1989 were introduced. In 1996, the Government of India (GoI) first realized the impending need to address the insufficient MSW management after numerous public interest litigations were filed against the GoI's inability in providing MSW management services to public. Based on the recommendations of an expert committee, the first Municipal Solid Waste (Management & Handling) Rules, 2000 were issued under the Environment Protection Act 1986. The Ministry of Environment, Forest and Climate Change (MoEF&CC) recently updated and issued the revised Solid Waste Management Rules 2016 by replacing the previous rules (MSWM Rules 2000; SWM 2016).

Within the new rules, clause 15(v)b recognizes RDF combustion as one of the Waste to Energy processes. Clause 21 provides criteria for waste to energy process as firstly non-recyclable waste having calorific value of 1500 kcal/kg (6.3 MJ/kg) or more shall not be disposed on landfills and shall only be utilised for generating energy either through refuse derived fuel or by giving away as feed stock for preparing refuse derived fuel. Secondly, wastes with a high calorific value shall be used for co-processing in cement or thermal power plants.

Clause (18) calls all industrial units using fuel and located within one hundred kilometres from a solid waste based RDF processing plant to make arrangements within 6 months from the date of notification of these rules to replace at least 5% of their fuel requirement by RDF.

Subsequently in 2016, the Indian Ministry of Urban Development (MoUD) released an Manual on MSWM for facilitating implementation of the new SWM rules 2016. The manual provides non-binding guidelines for "desirable RDF characteristics for co-processing in cement plants" as shown in Table 4.1. However, the manual does not specify methods that are permitted or required for the determination of characteristics or for quantification.

In 2016, draft guidelines for pre- and co-processing of hazardous and other wastes in cement plants were released by CPCB according to the Hazardous and Other Wastes (Management and Transboundary Movement) Rules, 2016 (CPCB 2017). These guidelines enable the cement industry to use different kinds of wastes, including hazardous and other wastes as substitute fuels and provide uniform operating procedures and emission standards for the pre- and co-processing of waste in cement kilns. However, these guidelines do not specify sampling procedures for the characterisation of the wastes. Currently, the CPHEEO has released

Table 4.1 Guidelines for RDF according to CPHEEO (2016)

Parameter	Limit value
Moisture content, MC	<20 wt.-%
Particle size (x_{50})	2D <120 mm
	3D <70 mm
Net Calorific Value, NCV	>3000 kcal/kg (>12.6 MJ/kg)
Chlorine	<0.7 wt.-%
Sulphur	<2 wt.-%
Other conditions – Free of restricted items (PVC, explosives, batteries, aerosol containers, biomedical wastes, etc.)	

a revised draft 'Guidelines on usage of RDF in various industries' with new standards to address the challenges related to usage of RDF in cement kilns such as heterogeneity of waste, varying demand for alternative fuels for co-processing, limited market development and compliance with the SWM rules 2016, (CPHEEO 2018). The guidelines propose quality criteria for RDF based on parameters listed in Table 4.2.

The cement kilns co-processing waste are obliged to follow the air emission standards specified in Hazardous and Other Wastes (Management and Transboundary Movement) Rules, 2016 (H&OW (M&TBM) 2016). For pre- and co-processing of wastes, all cement kilns and pre-processing facilities have to follow the standard operating procedures (SOP) specified by CPCB (2017). Apart from operational guidelines, the SOP demands products with chlorine content >1.5 wt.% and other toxic components to be injected in to main burner of cement kilns. Additionally, waste derived fuels shall not be used in cement kilns when the temperature of the calciner or the main burner is below 850 °C and 1100 °C respectively.

Although no binding documents for definition of permissible input materials or the utilisation of RDF are available, the emission standards for incinerators or thermal technologies in solid waste treatment or disposal facility in India are defined by the SWM rules 2016 (SWM 2016) listed in Table 4.3.

Table 4.4 shows a comparison of emissions standards for incineration of MSW specified in the existing MSW Rules 2016 with the standards of European Union and Germany (Dube et al. 2013).

However, the composition of Indian mixed waste is currently not comparable with residual waste from industrialised countries such as Germany. Dube et al. (2013) mentioned that due to a generally lower income situation, the use of goods containing chlorine, such as PVC, is significantly lower and so is the chlorine content of the waste. Besides, the informal waste pickers select PVC very thoroughly out of the collected waste. Nevertheless, emissions of highly toxic compounds such as dioxins and furans cannot be completely ruled out.

Table 4.2 Proposed standards for segregated combustible fraction (SCF) and RDF (CPHEEO 2018)

Parameters	SCF	RDF – Grade III	RDF – Grade II	RDF – Grade I
Intended use	Input material for the Waste to Energy plant or RDF pre-processing facility	For co-processing directly or after processing with other waste materials in cement kiln	For direct co-processing in cement kiln	For direct co-processing in cement kiln
Anything above 400 mm has to be mutually agreed between Urban Local Body/ SCF Supplier and Cement Plants.	Size	$\leq$50 mm or $\leq$ 20 mm depending upon use in ILC or SLC, respectively		
Ash – maximum permissible	$\leq$20 wt.-%[a]	$\leq$15 wt.-%	$\leq$10 wt.-%	$\leq$10 wt.-%
Moisture – maximum permissible	$\leq$35 wt.-%	$\leq$ 20 wt.-%	$\leq$15 wt.-%	$\leq$10 wt.-%
Chlorine – maximum permissible	$\leq$ 1.0 wt.-%[a]	$\leq$ 1.0 wt.-%	$\leq$ 0.7 wt.-%	$\leq$ 0.5 wt.-%
Sulphur – maximum permissible $\leq$1.5 wt.-%[a]			$\leq$1.5 wt.-%	
Net Calorific Value (NCV)[b]	$\geq$1500 kcal/kg	$\geq$3000 kcal/kg	$\geq$3750 kcal/kg	$\geq$4500 kcal/kg

[a]If blending process is done in cement plants, the deviations in recommitted limit for ash, chlorine and sulphur content can be mutually agreed between urban local body/SCF supplier and cement plants

[b]Range of variations with acceptable in NCV can be mutually decided between RDF manufacturer and cement plants

Table 4.3 Emission limits for cement kilns co-processing in India (CPCB 2015; SWM 2016)

Parameter	Unit		Detection
Totat dust	[mg/m^3]	$\leq$50 (or 0.125 kg/t of clinker)	Continuous
HCl	[mg/m^3]	10	Continuous
HF	[mg/m^3]	1	Continuous
NO_x	[mg/m^3]	800[a]	Continuous
		600[b]	Continuous
Hg, Cd, Th and their compounds	[mg/m^3]	0.05	By analysis
Sb, As, Pb, Cr, Co, Cu, Mn, Ni, V	[mg/m^3]	0.5	By analysis
Dioxins and Furans	[ngl-TEQ/m^3]	0.1	By analysis
SO_2	[mg/m^3]	100[c]	Continuous

[a]Plants commissioned before 1.8.2015

[b]Plants commissioned after 1.8.2015

[c]Relaxable up to 400 by SPCBs in special cases, CPCB proposed 100 (for <0.5% sulphur in raw materials), 1000 (for >0.5% sulphur in raw materials); (Units at 760 mm Hg, dry air, 10% oxygen, 273 K)

Table 4.4 Comparison of emission standards for incineration

Parameter	Units	MSW Rules 2016 (India)	Directive 2010/ 76/EC (Europe)[a]	17. BImSchV (Germany)[b]
Particulates/ SPM	[mg/m^3]	50	30	10
HCl	[mg/m^3]	50	10	10
SO_2	[mg/m^3]	200	50	50
CO	[mg/m^3]	50/100		50
Total Organic Carbon (TOC)	[mg/m^3]	20	10	
HF	[mg/m^3]	4	1	1
NO_x	[mg/m^3]	400	500/800	200
Total dioxins and furans	[ng TEQ/ m^3]	0.1	0.1	0.1
Cd, Th and their compounds	[mg/m^3]	0.05	0.05	0.05
Hg and its compounds	[mg/m^3]	0.05	0.05	0.03
Sb, As, Pb, Cr, Co, Cu, Mn, Ni, V and their compounds	[mg/m^3]	0.5	0.5	0.5

Note: All values corrected to 11% oxygen on a dry basis
[a]Directive 2010/75/EU of the European Parliament and of the Council of 24 November 2010 on industrial emissions (integrated pollution prevention and control)
[b]17.BImSchV-Ordinance (2009) on the Incineration and Co-incineration of Waste
Analysis of half hourly average are required

3.2 *Legislation and Standards in Europe and Germany*

Experiences of other countries on RDF legislation may support the utilisation of RDF in India and the development of national quality standards. In Europe, the production and utilisation of RDF reaches back to the 1970s. After unsolved technical problems with pollutants within the RDF during the 1980s and 1990 and the subsequent closure of RDF plants in Europe and the United States (Rotter 2010; Velis et al. 2010), a stronger focus was set on market and quality requirements for the production of RDF in contemporary SWM. In 1999, the production of RDF was mostly affected by the European Landfill Directive (1999/31/EC), which mandated the EU member states to utilize waste in higher levels of the waste hierarchy (e.g. recycling, energy recovery) than landfill. Only treated, non-valuable waste with minor energy content are allowed to be landfilled, while non-recyclable material with high calorific value has been shifted to thermal treatment plants or to plants for the production of RDF. According to EC law, RDF retains its waste status after processing and remains subject to the European waste licensing system.

Italy introduced 14 national standardised document on RDF (UNI 9903:1–14) and separated it further into RDF (with NCV $\geq$15 MJ/kg, MC $\leq$25 wt.-%) and RDF with high quality (RDF_Q) (with NCV $\geq$20 MJ/kg, MC $\leq$18 wt.-%), in which only

RDF_Q was considered a renewable energy source depending on its content of biodegradable waste.

To differentiate waste derived fuel products with tighter quality requirements from the broader and varying definition of RDF, the term solid recovered fuels (SRF) was introduced. In order to define harmonised technical and quality standards for RDF in Europe the Technical Committee on SRF of the European Committee for Standardisation (CEN) was established in 2003. As an outcome of the committee's work, the European Standard EN 15359:2011 Solid recovered fuels-specifications and classes were released, specifying a classification system for waste derived solid fuels. According to Ragazzi and Rada (2012), the introduction of SRF with defined quality characteristics provides information on RDF in further detail. As per the definition, the main objective of the production of SRF is to maximise the efficiency of energy recovery.

The standardised document's main goal is to improve the European trade of RDF/SRF, to increase its acceptance in the European fuel market and to facilitate its monitoring, utilization and approval. SRF is produced out of non-hazardous waste; potential sources may be production of specific waste, MSW, industrial waste, commercial waste, construction and demolition waste and sewage sludge. The processing of SRF is visualised in Fig. 4.1.

The classification system is essentially based on fuel characteristics, in particular NCV, chlorine content and mercury content. Within this classification, the NCV represents economic concerns, the chlorine content technical aspects, and mercury content an environmental parameter. Each parameter is separated into five classes, while the three parameters are of equal relevance for the final classification. The final values of NCV and chlorine content are determined as arithmetic mean value, the mercury content in addition as 80th percentile (of which the higher of both value determines the classification for mercury). The limit values and classes are presented in Table 4.5.

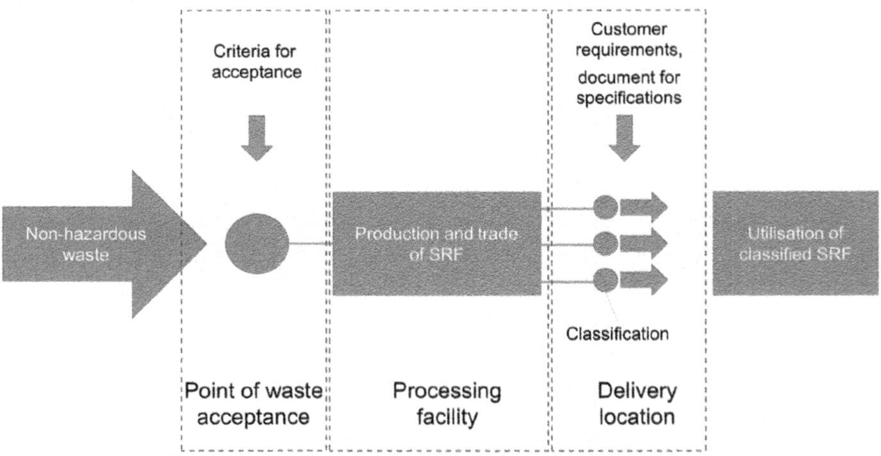

Fig. 4.1 Waste processing and classification of SRF according to EN 15359

Table 4.5 Classification standards of SRF according to EN 15359

Classification parameter	Statistical measure	Unit	Classes				
			1	2	3	4	5
NCV	Mean	[MJ/kg][a]	≥ 25	≥ 20	≥ 15	≥ 10	≥ 3
Chlorine	Mean	[wt.-%][b]	≤ 0.2	≤ 0.6	≤ 1.0	≤ 1.5	≤ 3.0
Mercury	Median	[mg/MJ][a]	≤ 0.02	≤ 0.03	≤ 0.08	≤ 0.15	≤ 0.50
	80th percentile	[mg/MJ][a]	≤ 0.04	≤ 0.06	≤ 0.16	≤ 0.30	≤ 1.00

[a]As delivered
[b]Water-free basis

The specifications for the SRF have been noted into a classification document. Beside the classification according to Table 4.5, the following additional information on the SRF characteristics are mandatory:

- Type of waste used for processing SRF
- Form of particles (e.g. pellets, powder, balls)
- Size of particles and its distribution
- Water and ash content
- Individual and total heavy metal composition (As, Cd, Cr, Co, Cu, Hg, Mn, Ni, Pb, Sb, Tl, Vn)

Further parameters like biomass content or bulk density can be determined on a voluntary basis. In order to ensure comparable values, additional standardised documents were created for the sampling and analysis of RDF/SRF (Sarc and Lorber 2013). The main sampling and analytical standards are described in EN 15400–15415 and EN 15440–15442. The standards define the analysis of, among others, the NCV, ash and moisture content, procedures for the determination of sulphur, chlorine, fluorine and bromine, as well as for the particle size distribution and sample collection. All standards of the CEN/TC343 have to be implemented by the EU member states. The classification process of the EU enables plant operators to acquire only RDF classes which comply with their specific technical plant configuration, approved emissions values and potential substrate mixing ratios.

In Germany, an additional certification for RDF/SRF (RAL-GZ 724) with quality standards in further detail was developed in 2002 to ensure continuous product qualities, limitations of heavy metal values, fixed sampling and analysis procedures and an improved monitoring of own and external product qualities. The quality certification does not conflict with the standards given in the EN 15359, and extends the European standard by defined limit values for heavy metals. The products certified after RAL-GZ 724 is traded as "Secondary Fuel" and granted trademark protection. The certification standard differentiates the combustible solid fraction of non-hazardous waste into a high-calorific fraction and secondary fuels. The high-calorific fraction is produced by only minor processing steps and shows a lower preparation depth (e.g. larger particle) than secondary fuels. Secondary fuels have to have a particle size <30 mm and must lie within the limit values for heavy metals shown in Table 4.6.

Table 4.6 Heavy metal standards for secondary fuel according to RAL-GZ 724

Parameter	Median value	80th percentile value
	[mg/MJ DS]	
As	0.31	0.81
Cd	0.25	0.56
Co	0.38	0.75
Cr	7.8	16.0
Hg	0.038	0.075
Mn	16	31
Ni	5.0	10.0
Pb	12	25
Sb	3.1	7.5
Sn	1.9	4.4
Tl	0.063	0.130
Vn	0.63	1.60

NCV, moisture, ash and chlorine content have to be documented. Since these parameters are considered as dependent in the respective process technology of the incineration plant with varying impacts from case to case, no limit values are given in the standard. The sampling process has to be done following guideline LAGA PN98 (LAGA 2019). For a produced amount of <20,000 tons per year, one individual sample has to be collected for every 10 tons, while 25 individual samples form one mixed sample for analysis. In order to ensure adequate RDF quality throughout, analysis by certified external laboratories and experts and an internal self-control mechanism are important.

4 Current Situation on RDF Generation and Use

4.1 General Overview

According to Bilitewski and Härdtle (2013), main targets of thermal treatment are the inertisation of residual wastes with minimal emissions, the destruction of organic toxins as well as the concentration of inorganic toxins, and volume reduction of the waste for disposal. In a sustainable waste management system, energy recovery and the production of substitute fuels are therefore only secondary objectives and reserved for non-hazardous waste.

The utilization of RDF in industrial applications offers a higher flexibility than waste incineration processes due to lower investment costs and offers a higher flexibility for future waste management and recycling concepts (EC 2003). Ensured that emission standards are met, the substitution of fossil fuels (e.g. coal, oil) with RDF in cement kilns offers great environmental benefits compared to its utilization in waste incineration plants (EC 2003). According to Parlikar (2017), the power generation through waste incineration provides an energy recovery rate of <30% and

0% of material recovery, while processing the high calorific fraction of MSW to RDF and utilization in cement kilns achieves >80% energy recovery and 100% material recovery.

The technologies of RDF utilization include spreader stroker fired boilers, suspension fired boilers, fluidized bed units and cyclone furnace units (CPHEEO 2018). The higher combustion temperatures of up to 1450 °C and the longer residence times of cement kilns when compared to waste incineration processes lead to a higher destruction rate of hazardous substances in RDF and the integration of the inorganic RDF-materials in the clinker (CPCB 2017). As a result, no incineration ash of RDF has to be disposed during its utilization in cement kilns. Additionally, acidic air emissions are neutralized by alkaline atmosphere within the combustion chamber, which subsequently reduces flue gas cleaning efforts (CPCB 2017).

An example of cement kiln with cyclone preheater, calciner and pre-combustion chamber is presented in Fig. 4.2. Cement kilns are able to utilize different quality types of RDF in different process steps – the pre-calciner, the main burner at the rotary kiln outlet or at rotary kiln inlet end (Baier 2016). Key parameters for the selection of the RDF injection point are calorific value and particle size. IFC (2017), Dembla (2017) and Parlikar (2017) describe suitable calorific values of RDF for utilization in cement kilns within a range of 11.7–20.9 MJ/kg. Usually, products to be used in the pre-calciner require a calorific value of 12–17 MJ/kg and a particle

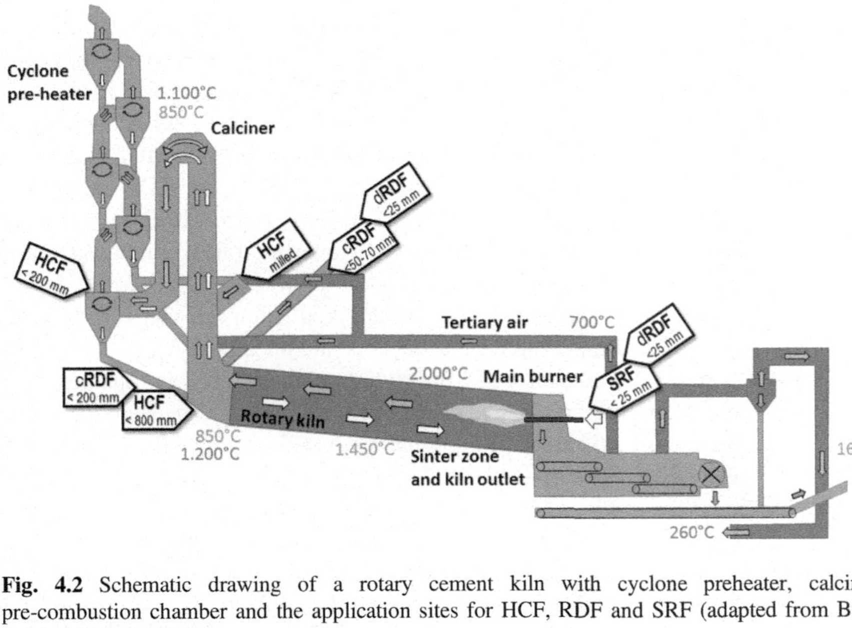

Fig. 4.2 Schematic drawing of a rotary cement kiln with cyclone preheater, calciner, pre-combustion chamber and the application sites for HCF, RDF and SRF (adapted from Baier 2016; BREF-WI 2017)

size of 50–80 mm (IFC 2017). However, for HCF or RDF to be added, special input aggregates must be installed on the calciner (Baier 2016).

The generation of RDF for the pre-calciner requires further processing; 65–75 wt.-% of the input material will achieve the required quality and 25–35 wt.-% will remain as product. RDF for the main burner has high demands regarding product quality and energy content. The required calorific value ranges between 18–23 MJ/kg with particle size of 20–35 mm (IFC 2017). To meet the quality standards, large shares of commercial and industrial waste or residues from waste recycling are commonly required (IFC 2017).

4.2 *Status in India*

According to CPCB (2017), 65 million tons of MSW per year is generated in India, of which 15–20 wt.-% can be classified as non-recyclable, segregated combustible fraction (SCF), which can be processed further to RDF. In 2017, 138,700 tons per day of Indian urban MSW were generated within a radius of 200 km of cement kilns, of which approx. 22.5% may be converted into RDF (Swami 2017). The Indian cement demand is projected to amount 500–640 million tons per year in 2025, resulting in a rising need for fuels (Kannan 2017).

In India, one current example is the MSW treatment facility in Goa, which is sending 30 tons per day of RDF to the ACC Wadi cement kiln, approx. 470 km away, for co-processing (SFC Environmental Technology 2017a, b; Swami 2017). Additional trials with RDF in cement kilns in India have been done so far for different waste streams. Out of 29 RDF plants in 2017, 'few success stories' could be reported (Swami 2017). In 2010, 23 RDF processing plants were operational, producing 24.12 wt.-% RDF out of 9720 tons of MSW per day (GIZ-IGEP 2013). In the absence of clear quality standards, the market prices for RDF with GCV 3200–3800 kcal/kg (13.4–15.9 MJ/kg) range widely from 420–3000 INR per tonne (Michaelowa et al. 2015; GIZ-IGEP 2013). However, if RDF quality standards were agreed, market forces could govern the actual prices.

In practice, there are several challenges for the production and recovery of RDF. Potential barriers, according to the RDF producers, are uncertainty about the acceptance in RDF, the risk of rejection and uncertainty about supply and quality (high humidity, inconsistent particle size and calorific value). The heterogeneity of wastes and an inadequate infrastructure are further major obstacles to the non-acceptance of RDF (GIZ-IGEP 2013; Sharma 2017; CPHEEO 2018). According to Parlikar (2017), the currently produced RDF fractions do not achieve the product quality required for co-processing in cement kilns, as national quality standards for RDF are still lacking. Currently, there are no product limitations for ash, water, sulphur, chloride and heavy metal contents. Hence, the produced RDF may only be used in a waste incineration process (Parlikar 2017).

4.3 Status in EU and in Germany

As a contrast, Germany's experience with waste incineration, energy recovery and RDF utilization reaches back to the 1970s. In 2014, approx. 47.5 million tons of waste were being treated thermally, of which 46% were treated within 66 waste incineration/waste to energy plants and 28% used as fuel substitute in combustion process (e.g. RDF plant, coal power plants, co-combustion in cement kilns or steel mills) (Hiebel et al. 2017; Flamme et al. 2018). Out of the 66 thermal plants covering 1.53% of the annual German primary energy demand in 2017, 60 plants are equipped with a CHP, while six plants are only generating power (Flamme et al. 2018; UBA 2018). 32 power plants using only RDF have a total national capacity of 6.3 Mio. tons per year (Flamme et al. 2018). Apart from waste/RDF to energy plants, 34 cement kilns and two lime plants are approved for RDF co-processing (Flamme et al. 2018). Also, 25 coal power plants had an approval to co-process waste in 2016, treating approx. 1.5 million tons of waste per year (Birnstengel et al. 2018). Out of 3.2 Mio. tons of waste derived fuels used in 34 cement kilns in 2015 (Flamme et al. 2018; Birnstengel et al. 2018), 10% was segregated combustible MSW as well 21% plastic fractions and 2.9% paper fractions of industrial and commercial waste (VDZ 2016a). The thermal substitution rate of German cement kilns with alternative fuels amounted 64.6% in 2015 (VDZ 2016b). In addition, 16 coal-fired power plants co-incinerated waste amounting to around 1.5 million tons. The types of waste used range from secondary fuels (eight coal-fired power plants) to paper and fibre sludge, sewage sludge, plastics and foils, spit, animal meal, hazardous waste and organic liquids (Birnstengel et al. 2018).

Although Germany has high standards for recycling and a strong agenda to introduce circular economy, 53% of 5.92 Mio. tons of separately collected plastic wastes were used in a thermal process in 2015, in which 1.09 Mio. tons were processed to and used as RDF (Hiebel et al. 2017). Hazardous substances or additives within large shares of plastics, as well as complex material mixtures and composites, improve the properties of plastic products, but complicate or impede their subsequent recycling process. Although newly introduced requirements on product properties and recyclability are slowly changing the properties of plastic waste, durable products made of non-recyclable plastics will remain in the anthropogenic reservoir and continue to enter the waste management system throughout the next decades. Hence, thermal treatment as well as processing of RDF is considered to remain a vital part in German waste management during the transition process towards a circular economy, in order to inertise waste material and concentrate hazardous compounds.

Through comparative evaluation study Vodegel and Davidovic (2017) found that different types of power plants and production plants when using RDF have different average energetic efficiencies (fuel utilisation rates). Due to the direct use in the cement clinker burning process, the net efficiencies of RDF use in cement kilns are over 70%, just as high as with the use of substituted primary fuels such as lignite or hard coal. The net efficiencies of RDF power plants depend strongly on the useful energies produced. For example, the net efficiency of plants with full power generation is only approx. 20%. With pure process steam output, however, already at approx. 80%; with combined heat and power between 30 and over 70%. In Germany waste incineration plants achieved net efficiencies of RDF use of 12% (old plants) to over 20% (new plants). With pure steam output, e.g. directly to a power plant, the efficiency is almost 80%, with CHP use depending on the heat output between approx. 20 and 70%.

5 RDF Processing

5.1 Types of RDF

As stated in the definition, RDF is produced by mechanically separating the combustible fraction from the non-combustible fraction of solid waste. RDF can thus be produced from municipal solid waste (MSW) through a number of different processes that includes separation at source; sorting or mechanical separation; size reduction (shredding, chipping and milling); separation and screening; blending; drying and pelletising; packaging; and storage (EC 2003 2010).

There are two basic RDF processes, each producing a distinctive product, known as densified RDF (dRDF) and coarse RDF (cRDF) (high calorific fraction) respectively (McDougall et al. 2003). Depending on market needs dRDF is produced in the form of pellets or as loose fill with fluffy consistency. dRDF is dried before transport or pelletising, which improves its stability and transportability. Also, storage and handling is similar to other solid fuels. It can either be burned alone, or co-fired with coal or other solid fuels. dRDF requires considerable processing, including drying and pelletizing, and so has a relatively high processing energy requirement.

As a result, there has been interest in the alternative cRDF (high calorific fraction) recently. This comes in the form of a coarsely shredded product. cRDF requires less processing, but it cannot be stored for long periods if it has not been dried. There is a danger that anaerobic conversion processes that release methane will also take place during interim storage due to microbiological activity. With a volume fraction of 4.4 vol.-% in the air (lower explosion limit) and 16.5 vol.-% in the air (upper explosion limit) it forms explosive mixtures or dangerous atmospheres (Weichgrebe 2015).

In practice, a distinction is made between high calorific fraction (HCF), refused derived fuel (RDF) and solid recovered fuel (SRF) in Germany. Thereby HCF represents the separated combustible material (<300 mm, NVC < 9 MJ/kg) which can be used directly in so-called pre-combustion chambers or is processed directly or mixed with commercial and industrial waste to solid fuels; RDF represents calcinator fuel which is either prepared for use in cement kilns (<60–120 mm, NCV 15–19 MJ/kg) or is used with other wastes as starting material for the preparation to main burner fuel. (A subcategory are chips from shredded used tyres (<60–120 mm, NCV 28 MJ/kg).) SRF represents fuel for the main burner which is processed so well by mixing, shredding and sifting that it meets the high requirements (e.g. fast burnout) of the main burner (<15–30 mm, NCV >20–28 MJ/kg).

Several technologies are currently established for the production of a high caloric fraction from MSW, which can be processed as RDF, i.e. mechanical treatment (MT) alone, mechanical biological treatment (MBT), mechanical biological stabilisation (MBS) process, and mechanical physical stabilisation (MPS) (EC 2003). In a mechanical biological pre-treatment plant (MBT), metals and inert are separated out and organic fractions are screened out for further stabilisation using composting processes, either with or without a digestion phase.

A "dry stabilisation process", in which residual waste (excluding inerts and metals) is effectively dried and stabilised by a composting process, can also produce RDF. The heat is supplied directly from the exothermic aerobic microbial degradation process (bio-drying).

However, the process must be controlled so that self-ignition does not occur. The remaining product is a residual mass with a higher calorific value that is directly suitable for combustion. Nevertheless, ferrous and non-ferrous metals are still recovered from the material flow after drying, e.g. by over belt magnets or eddy-current devices.

5.2 *Stages of RDF Production*

Details of the processing line in different RDF plants vary. However, the basic RDF process can be broken down into five distinct stages, i.e. waste reception and storage, waste liberation and screening, fuel refining, fuel preparation, and fuel storage and quality control (GECF 2016; Beckmann et al. 2012; Sabery 2004). A generalised RDF production steps are shown in Fig. 4.3.

As the first step, mixed MSW is delivered by the collection vehicles onto a tipping floor, where any desired bulk items are removed. This initial short-term storage stage acts as a buffer to provide the RDF production process with a steady feedstock level (Sabery 2004).

Secondly, the waste is transported via a conveyor belt to be shredded, e.g. in two-roller shredders, crushers or mills, and then screened, e.g. with a rotary drum screen or a ballistic bar sizer. It performs three functions – completing the bag emptying process, removing the undersize (fines, <60 mm) fractions, and separating out the oversize materials from the fuel fraction (GECF 2016; Beckmann et al. 2012; Sabery 2004). The fines fraction contains the high moisture content organic and putrescible material, as well as ash, dust, sand, broken glass and other inert materials. The wet organic materials (nearly 4–60 wt.%) can then undergo further treatment such as composting or anaerobic digestion, and can be used as a soil conditioner for landfill restoration work or be landfilled. In some cases, the putrescible fraction is kept in place to enable the mass of material to be dried through biological treatment (the process of 'dry stabilisation'). The oversize fraction contains mainly of large pieces of paper, board, and plastic film, and is usually landfilled along with other

production steps are shown in

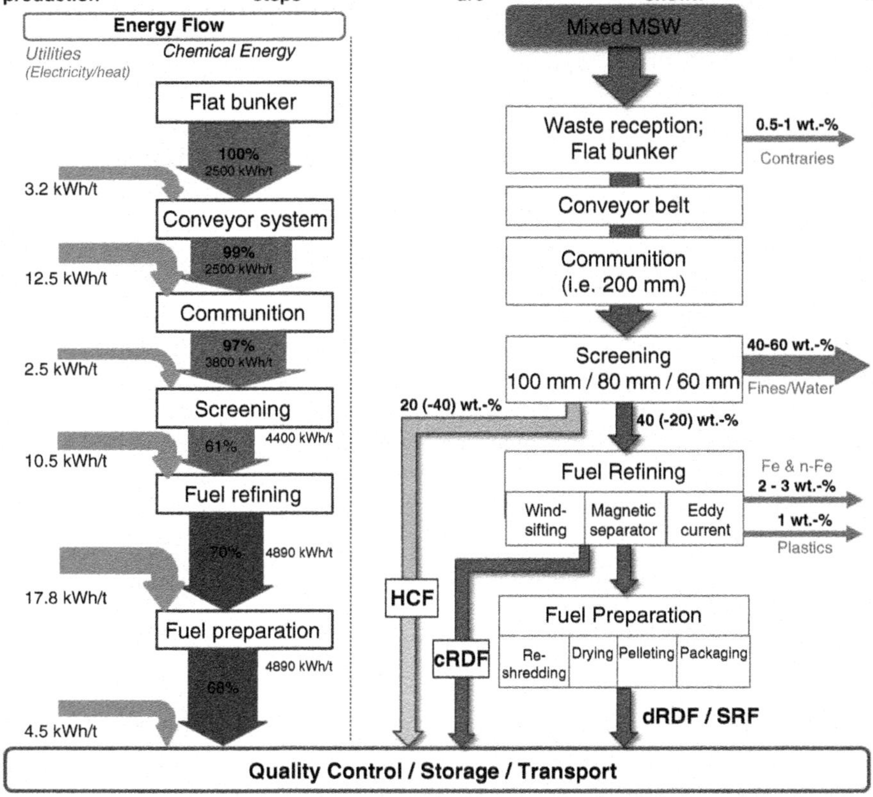

Fig. 4.3 Stages of RDF production from MSW including energy flow

residues. The remaining fraction produced by this stage can be used as high calorific fraction, though it still contains metals and other non-combustible materials (Beckmann et al. 2012; Sabery 2004).

Thirdly, the medium fraction is refined that involves size reduction, classification, and magnetic separation. Size reduction using a shredder or hammer mill aids the separation into light and dense fractions. The density separation (classification) stage is necessary to separate the heavy fraction (metals, dense plastics) from the combustible light fraction (paper, plastic film), which will go onto form the densified RDF product. Two main methods are used to achieve this: air classification, and ballistic separation (Ketelsen et al. 2011). For the recovery of ferrous metals (Fe) overband magnets are usually installed and for the recovery of non-ferrous metals (nFe), in particular aluminium, eddy current separation is used (2–3% by weight). The light fraction, together with the remains of the magnetically sorted heavy fraction, can be used as a more refined form of coarse RDF (EC 2003; Sabery 2004).

Table 4.7 Assembled RDF varieties (Lingk 2015)

Parameter	Type	cRDF	cRDF	dRDF	dRDF
		Fluffy	Rolled	Soft pellet	Hard pellet
Particle size	[mm]	0–45	0–22	0–22	0–22
Bulk density	[kg/m^3]	130	250–300	350–400	650

Fourthly, the fuel preparation step is performed through the conversion of the fuel-rich fraction into a dry, dense pellet form by re-shredding, drying, and then pelletising. Secondary shredding is needed to reduce the particle size of the fuel fraction to the size needed for the pelletising operation, and drying reduces the moisture content from about 30 wt.-% to around 12 wt.-% (Beckmann et al. 2012). The used dryers are basic pneumatic conveying systems that operate on hot combustion gas from natural gas burners. Once the combustible fraction is dry, organic and inert residues can easily be screened out, reducing the ash content of the product. Most of the chlorine, heavy metal, and silicates in the product are contained within this inert residue. After this stage densified RDF can be produced with a final ash content of 10 wt.-% by weight and chlorine levels of 0.5 wt.-% (Beckmann et al. 2012; GECF 2016). In the absence of inert contaminants such as silicates, the calorific value of the material increases significantly. Densified RDF can either take the form of pellets or briquettes, though most plants use a pellet mill to densify the product. The densified RDF can be stored more easily after drying and pelletisation, but its temperature has to be monitored by infrared cameras. Assembled RDF is usually marketed in four categories:as shown in Table 4.7.

5.3 Energy Consumption

Several operations in the RDF process have significant electrical energy consumptions, in particular shredding/crushing (10.8 kWh/t), homogenisation (2.7 kWh/t), sieving/screening (0.5 kWh/t) and pelletising (9.5 kWh/t) (Sabery 2004; Ketelsen et al. 2011). Sabery (2004) reported energy consumption in other independent mechanical processing units, such as 0.4 kWh/t in Fe-separation, 1.1 kWh/t in wind shifting, 3.2 kWh/t for conveyor system and 0.2 kWh/t for bale pressing. The overall electrical energy consumption for HCF has been estimated as 23 kWh per tons of annual plant input (Ketelsen 2013) and the RDF/SRF process has been estimated as 55.5 kWh/t of annual plant input. In addition, the drying process prior to pelletising requires around 400 MJ/t (input) heat energy (Gallardo et al. 2014). In plant where on-site combustion of RDF occurs, this drying heat requirement can be met by burning some of the produced RDF, or by using waste heat from the power generation system. Where no on-site burning of RDF occurs, heating by gas or other fuels will be needed.

According to Ketelsen et al. (2011), overall power consumption for MBT process with rotting is estimated at 59–63 kWh/t (input), and 30–50 kWh/t (input) for gas.

The power consumption of MBS and MPS plants, however, is 20–30 kWh/t (input) higher, while the gas consumption is almost equal (Ketelsen 2015).

Only mechanical processing requires 3.3 kWh energy per tons of input waste. Calorific value of mixed MSW is considered as 9.2 MJ/kg; however after screening, calorific value of light/coarse fraction rises to 10–12 MJ/kg. After fuel refining the calorific value can be further improved to 18 MJ/kg (Weichgrebe et al. 2016; Beckmann et al. 2012).

6 Potentials of RDF in India

6.1 Suitability of Waste Categories and Size Fractions for RDF Processing

In the following, the suitability of certain waste categories and size fractions within mixed MSW for RDF processing is presented for the case of Bangalore, India. Weichgrebe et al. (2017) and Speier et al. (2018a) described the methods and statistical standards followed for this comprehensive waste investigation including chemical analysis of different waste fractions. For the analysis of RDF potentials in mixed MSW, waste samples were obtained from household waste by door-to-door collection and from litter spot waste in West-zone Bangalore. The chemical composition of the waste samples was analysed, for evaluating the RDF suitability according to the quality criteria suggested for India by CPHEEO (2016). Each waste stream was sieved with 77 mm and 55 mm drum screen, and each mixed fraction was sorted manually into the fractions 'Organics', 'Paper', 'Plastics', 'Textiles', 'Composites', and 'Liquids'. The analysis results are presented in Table 4.8.

It is observed that in both collected waste streams the 'Organics' show the minimum required calorific value for thermal treatment of 1500 kcal/kg mandated by the SWM Rules 2016 (MoEFCC 2016). 'Plastics' in door-to-door collection stream and 'Composites' in both waste streams hold an NCV of 12.6 MJ/kg by CPHEEO (2016). For all other fractions, further treatment (e.g. drying) is required to achieve the quality criteria. As demonstrated by Speier et al. (2018a), wet organic waste shares in the MSW with low NCV decrease with increasing particle size. As a result, an increase of NCV is observed at larger particle sizes (Table 4.8). This emphasises the effectivity of size screening for the processing of RDF. The suggested sulphur content (<2 wt.-%) is maintained by all waste categories and sizes. Additionally, all categories and sizes are within the proposed chlorine limits of <0.7 wt.-%, except 'Liquids'. As mentioned in Table 4.3, one of the major sources of chlorines is found in kitchen waste due to the presence of chloride salts. The analysis showed that 'Liquids' were majorly comprised of packed, liquid food items (e.g. soups, gravies). Hence, the increased chlorine contents may be a result of high concentrations of food-related chloride salts. Elevated chlorine contents in

Table 4.8 Chemical composition related to RDF quality criteria exemplary of MSW samples from Bangalore (Weichgrebe et al. 2017)

		NCV		Sulphur [wt.-% of DS]	Chlorine [wt.-% of DS]	Mercury [mg/MJ]	Share [wt-% of total]
		[MJ/kg]	[kcal/kg]				
Door-to-door collected MSW	>77 mm	11.1	2664.45	0.4	0.0	0.001	32.2
	55–77 mm	8.8	2112.87	0.3	0.0	0.003	18.9
	14–55 mm	7.5	1802.10	0.4	0.1	0.001	39.0
	<14 mm	–	–	–	–	–	9.9
	Organics	4.7	1116.98	0.4	0.1	0.002	60.7
	Paper	9.9	2366.55	0.4	0.0	0.001	8.6
	Plastics	13.1	3138.82	0.3	0.0	0.001	9.8
	Textiles	12.2	2909.71	0.4	n.d.	0.002	4.1
	Composites	14.1	3370.46	0.2	0.0	0.000	2.9
	Others	–	–	–	–	–	13.9
Litter spot MSW	>77 mm	9.9	2360.85	0.3	0.1	0.002	35.7
	55–77 mm	4.8	1148.90	0.4	0.5	0.001	17.3
	14–55 mm	4.2	1013.23	0.4	0.4	0.001	35.5
	<14 mm	–	–	–	–	–	11.5
	Organics	4.2	1014.11	0.4	0.5	0.003	63.4
	Paper	8.0	1923.61	0.3	0.2	0.001	10.1
	Plastics	8.1	1932.31	0.3	0.1	0.001	11.1
	Textiles	9.6	2283.03	0.3	0.1	0.001	6.6
	Composites	13.5	3217.72	0.2	0.4	0.002	3.7
	Others	–	–	–	–	–	5.1

n.d. – not determined

'Organics' and waste $d < 55$ mm in litter spot MSW suggest a similar origin. When compared to the quality criteria for SRF given in EN 15359, the categories 'Plastics', 'Paper', 'Textiles' and 'Composites' as well as waste $d > 77$ mm are within quality classes 1 and 2 for mercury and chlorine, and classes 3 and 4 for NCV.

In summary, the fractions 'Plastics', 'Paper', 'Textiles' and 'Composites' of both collected and analysed MSW streams show a high suitability for the production of RDF. When compared to existing international quality standards like the EN 15359, the waste categories are able to achieve high quality levels. However, further processing is recommended to increase the NCV, and size screening i.e. by drum screen or by ballistic separator (BREF 2017).

6.2 *Potentials in MSW and Utilization Sectors*

Based on waste composition analyses published within the last decade in India, the RDF potentials out of MSW for Indian urban areas are calculated. The sorting categories used in the evaluated data sets were harmonised into eight main waste categories: organic, paper, plastic, glass, metal, inert materials, textiles and others. Of these data, paper, plastic, textiles and others (often containing composites, rubber, etc.), commonly with potentially suitable calorific values, are considered for the production of RDF. Additionally, a share of the organic fraction is assigned to the RDF fraction. Organic wastes usually show high moisture content with low calorific values. However, large organic waste items present in MSW (e.g. branches, coconut shells, wood) are potentially suitable for energy recovery. To account for these items, a share of 20 wt.-% of the total available organic waste is assumed as input for the combustible fraction. The estimation incorporates population growth, rising urbanization rates and an increasing per capita waste generation rate in India according to the World Bank (2015), Hoornweg and Bhada-Tata (2012) and CPHEEO (2016). For cities with more than one available waste composition data set, the mean value of the data is obtained.

The evaluation of 50 cites shows that in average, 31.4 wt.-% of the urban MSW would be suitable for the production of a high calorific fraction, with a strong deviation of $\sigma = 8.72$. This mean value is slightly higher than the projected RDF potentials within MSW described in Chap. 1 and lower than the value of 40 wt.-% given by CPHEEO (2018). Based on these findings, RDF potentials of Indian urban MSW consisting of organic 20 wt.-%, paper, plastics, textiles and others from MSW are projected in Table 4.9.

Speier et al. (2018a) investigated the composition of mixed MSW in Bangalore, India. The study reported a share of 22 wt.-% of valuable materials within the MSW plastic fraction, and share of 59 wt.-% potentially retrievable, valuable newspapers, cardboard and other papers within the MSW paper fraction. Considering recycling of these materials prior to the RDF processing, the average share of MSW for RDF processing is reduced to 25.6 wt.-%. Speier et al. (2018a) reported NCV values of individual fractions in the mixed MSW of Bangalore as follows: paper 10.2 MJ/kg, plastic 27.0 MJ/kg and textiles 14.7 MJ/kg. Speier et al. (2018a) reported that organic materials above a particle size of 55 mm mostly contain wood, coconut shells, tree trimmings or fibrous leaves. For the organic fraction within the RDF, the NCV of wood materials of approx. 18.5 MJ/kg is considered (Iacovidou et al. 2018), since the primary treatment step in common RDF plants includes size screening. Due to the heterogeneity of the category "other" waste, literature values were not applicable. Therefore, a calorific value of 11 MJ/kg is assumed for the following estimations. The total NCV is calculated as follows:

$$NCV_{hcf} = \frac{\sum NCV_i \cdot m_i}{m_{hcf}}$$

where i = organics, paper, plastics, textiles and others.

The projection revealed that the analysed waste fractions have an average calorific value of 4066 kcal/kg (17.0 MJ/kg) with $\sigma = 371$. This value is beyond the required minimum value of 1500 kcal/kg according to SWM Rules 2016 and 3000 kcal/kg recommended by CPHEEO (2016). Even the lowest value of 3039 kcal/kg remains within the recommended NCV range.

A comparative mass balance for RDF production from MSW through mechanical biological treatment (MBT) and mechanical biological stabilisation (MBS) processes are estimated for the case of West-Zone Bangalore, India with a population of 1.66 million. Considering an average waste generation of 0.43 kg/person/day (Speier et al. 2018a), the total mixed MSW is estimated to 260,537 tons of MSW per year. The detailed mass balance results for the MBT and MBS processes are presented in Fig. 4.4. Through the MBT process, approximately 10.8 wt.-% light RDF and 24.5 wt.-% medium RDF can be obtained from the mixed MSW. In comparison, MBS process can generate nearly 27.6 wt.-% RDF including 6.0 wt.-% light and 21.6 wt.-% medium RDF (Weichgrebe et al. 2016).

7 Conclusions

Sustainable management of Indian MSW currently poses a great challenge for urban infrastructures and potential risks for human health. According to the Indian waste hierarchy, waste which cannot be avoided, reused or recycled, but has suitable properties for combustion, has to be processed and converted to refuse derived fuel (RDF) which can be utilized as alternative fuel in various industries.

The analysis of compositional data of urban MSW from 50 Indian cities revealed that the average share of MSW available for RDF processing amounts 25.6 wt.-% with an average NCV of 4066 kcal/kg. When considering the annual MSW amount of 68.8 million tons, 48,254 tons per day of RDF would be available in India. An exemplary analysis of the combustible waste fraction in Bangalore further revealed that contents of critical parameters (i.e., chlorine, sulphur, mercury) are below the suggested limit values, and that the NCV of the MSW can be increased with less effort by size screening.

Therefore it is well understood that the production of RDF offers an attractive opportunity to both reduce the waste volumes and substitute fossil fuels, in various industries in India. However, policy and economic-market instruments such as pricing of waste disposal (fee from generators), quality parameters to utilize alternative fuels and technical capacity building of industries (to develop necessary infrastructure such as storage, handling, feeding mechanism, kiln design, etc.) are needed to enable RDF utilization. The new policy guidelines and ongoing consultation between the GOI and industries is expected to bring significant development in the coming years in Indian waste market.

In conclusion, the Indian MSW sector offers a great opportunity for the utilization of high quality RDF, substitution fossil fuels and improving both India's waste management system and CO_2 balance. Countries in EU show how efficient RDF

Table 4.9 RDF potentials of mixed solid waste collected in Indian cities (adapted after waste composition data, reference list in Appendix 1)

City	Population[a]	Pot. RDF amount tons/d	Pot. RDF ratio wt.-%	Segregated fractions of mixed Municipal Solid Waste: Organic (20%) tons/d	Paper tons/d	Plastics tons/d	Textiles tons/d	Others tons/d
Mumbai	12,478,447	1851.2	23.3	857.4	597.3	268.9	65.8	61.8
Delhi	11,007,835	1691.7	24.1	682.6	299.0	327.6	45.3	337.2
Bangalore	8,425,970	2086.5	38.8	640.6	550.7	585.3	124.7	185.3
Hyderabad	6,809,970	1234.1	28.4	437.1	492.7	160.8	80.3	63.2
Ahmedabad	5,570,585	915.1	25.8	340.9	109.9	181.0	165.8	117.5
Chennai	4,681,087	662.2	22.2	268.8	154.0	140.8	64.4	34.3
Lucknow	2,815,601	693.1	38.6	145.7	233.8	142.1	118.4	53.1
Kanpur	2,767,031	324.2	18.4	156.1	56.0	95.8	4.4	11.9
Nagpur	2,405,421	593.0	38.7	145.4	105.4	114.3	145.4	82.5
Indore	1,960,631	337.8	27.0	122.4	76.3	72.1	30.1	36.9
Bhopal	1,795,648	372.0	32.5	112.1	127.6	73.3	10.4	48.6
Visakhapatnam	1,730,320	254.7	23.1	108.1	79.8	51.0	13.3	2.6
Pimpri-Chinchwad	1,729,359	414.2	37.6	163.8	59.8	72.9	0.0	117.6
Patna	1,683,200	261.1	24.3	104.1	47.1	54.4	49.2	6.3
Vadodara	1,666,703	310.2	29.2	100.8	63.6	80.6	51.6	13.6
Agra	1,574,542	376.6	37.5	105.0	125.5	60.2	7.5	78.3
Nashik	1,486,973	334.6	35.3	77.9	53.0	82.2	102.1	19.5
Faridabad	1,404,653	334.7	37.4	75.3	76.8	123.0	37.1	22.6
Meerut	1,309,023	238.1	28.5	91.0	41.3	60.1	41.6	4.1
Rajkot	1,286,995	153.0	18.7	71.0	20.5	24.6	20.5	16.4
Varanasi	1,201,815	179.7	23.5	61.3	38.3	48.9	20.3	10.9
Dhanbad	1,161,561	217.2	29.3	69.5	53.3	41.2	32.7	20.5
Amritsar	1,132,761	166.8	23.1	88.7	27.0	22.2	0.0	28.9
Allahabad	1,117,094	241.2	33.9	50.6	51.8	73.6	52.3	13.0

Ranchi	1,073,440	159.7	23.3	70.5	21.7	23.6	34.0	9.9
Jabalpur	1,054,336	216.2	32.2	64.6	51.6	55.8	29.7	14.5
Raipur	1,010,087	199.9	31.0	66.2	53.5	45.5	25.1	9.5
Guwahati	963,429	241.0	39.2	61.9	90.6	81.3	6.7	0.5
Mysore	887,446	117.7	20.8	47.0	0.0	56.6	8.5	5.7
Jalandhar	862,196	167.7	30.5	40.7	35.7	17.6	2.7	70.9
Bhuwaneshwar	837,737	142.7	26.7	53.2	30.7	30.4	17.1	11.2
Thiruvananthapuram	752,490	148.7	31.0	59.6	36.0	35.5	4.2	13.4
Jamshedpur	629,659	119.2	29.7	34.8	41.1	21.2	12.0	10.1
Warangal	620,116	121.7	30.8	31.5	33.3	21.1	18.1	17.8
Kochi	601,574	86.8	22.6	46.5	9.4	2.1	3.8	24.9
Dehradun	578,420	127.7	34.6	48.5	32.4	28.7	10.3	7.8
Asansol	564,491	97.3	27.0	36.2	38.4	10.0	11.0	1.7
Jalgaon	460,468	67.7	23.1	36.2	13.9	9.8	7.1	0.7
Agartala	399,688	69.4	27.2	30.0	20.7	11.3	5.5	1.9
Kottayam	357,533	77.3	33.9	34.2	19.8	11.2	3.4	8.7
Kohima	267,988	58.6	34.3	20.6	17.2	12.8	2.9	5.1
Pondicherry	241,773	54.3	35.2	15.8	16.1	11.1	0.4	10.9
Gandhinagar	208,299	37.0	27.9	9.1	7.4	8.5	7.0	4.9
Daman	191,173	40.1	32.9	7.2	12.8	10.9	6.0	3.2
Cuddalore	173,361	40.0	36.2	12.9	10.1	3.9	2.8	10.2
Palakkad	131,019	51.6	61.7	2.5	21.4	5.3	0.6	21.8
Sangamner	61,958	18.6	47.0	3.3	2.4	2.4	1.0	9.5
Dharapuram	56,163	15.0	42.0	3.8	1.2	6.3	2.6	1.2

[a]Census of India (2011)

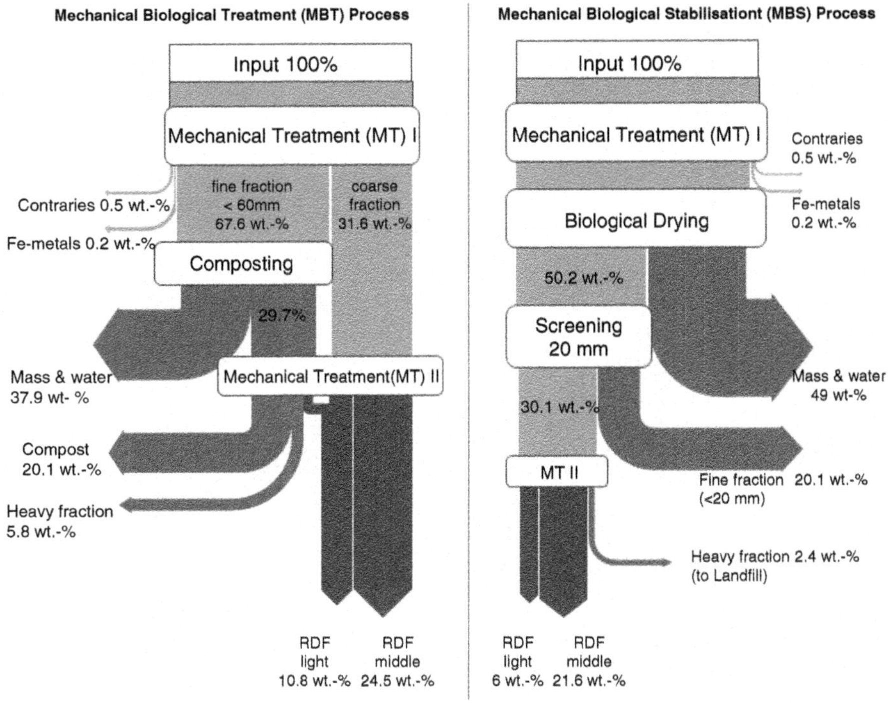

Fig. 4.4 Calculated mass balance of RDF production from MSW through MBT and MBS processes on the example of Bangalore. (Modified after Weichgrebe et al. 2016)

may be used as substitute for fossil fuel as co-combustion in cement kilns, coal-power plants, or in mono-combustion plants. It is now the challenge to develop the value added chain of waste to energy of mixed solid waste under the framework of the agreed waste hierarchy. The amount and composition of waste in India will change due to the increase of population and GDP, the change in consumer behaviour but also the efforts in waste segregation and recycling. Thus, defined standards for RDF categories, qualities, operation and application, are essential to establish RDF utilization routes and plants, and to ensure their economic viability while guaranteeing environmentally sound RDF use. However, the cement industry, the main recipient of the RDF, is not only exposed to the pressure of legislation, but also within the industry. The local plant operator—whether independent or affiliated—only has individual scope for remedial action. In order to determine the maximum absorption capacity and make optimum use of it, infrastructural and construction measures are necessary. Due to the high level of investment and the consideration of environmental protection concerns, long-term planning horizons, strategies and security as well as the cooperation of RDF producers and customers, ULBs, NGOs and regulatory authorities are indispensable for all parties involved.

References

AGPU (2018). Everything about PVC Recycling, AGPU e.V. https://www.agpu.de/en/all-about-pvc/recycling. Accessed 23/10/2018.

Banerjee, P. (2016). Gone to waste: How India is drowning in garbage. Hindustan Times, 09 February. http://www.hindustantimes.com/india/india-s-cities-are-faced-with-a-severewaste-management-crisis/story-vk1Qs9PJT8l1bPLCJKsOTP.html. Accessed 31/05/2018

Baier, H. (2016). RDF usage before, during and after the exit of coal combustion. ITAD Conference, Würzburg. Sept. 20th, 2016.

Beckmann, M. and Ncube, S. (2007). Characterisation of Refuse Derived Fuels (RDF) in reference to the Fuel Technical Properties. *In:* Proceedings of the International Conference on Incineration and Thermal Treatment Technologies – IT3. Phoenix, USA. ISBN: 9780923204822.

Beckmann, M., Pohl, M., Bernhard, D., Gebauer, K. (2012). Criteria for solid recovered fuels as a substitute for fossil fuels - a review. In: Waste Management & Research. April 2012, vol. 30, S. 354–369. doi:https://doi.org/10.1177/0734242x12441237

Bilitewski, B. and Härdtle, G. (2013). Abfallwirtschaft – Handbuch für Praxis und Lehre. 4th edition. Springer, Berlin and Heidelberg. https://doi.org/10.1007/978-3-540-79531-5

BImSchV (2009) Seventeenth Ordinance on the Implementation of the Federal Immission Control Act (Ordinance on the Incineration and Co-incineration of Waste - 17th BImSchV); version published on 14 August 2003 (Federal Law Gazette BGBL. I p. 1633), as amended by Article 2 of the Ordinance of 27 January 2009 (Federal Law Gazette I p. 129).

Birnstengel, B., Eckhardt, M., Häusler, A., Hoffmeister, J., Labinsky, A., Lambert, J., Lühr, O., Schütz, N., Simpson, R., Becker, G., Gellenbeck, K. and Weppel, J. (2018). Statusbericht der deutschen Kreislaufwirtschaft 2018. Einblicke und Aussichten. https://www.vku.de/fileadmin/user_upload/Verbandsseite/Themen/Europa/Statusbericht_2018_Ansicht_und_Druck.pdf. Accessed 23/10/2018.

BREF-WI (2017). Waste Incineration; Best available techniques. Reference document (BREFs) developed under the IPPC Directive and the IED, Draft May 2017.

Census of India (2011). Government of India, Ministry of Home Affairs, Office of the Registrar General & Census Commissioner, India. Population Enumeration Data (Final Population). http://www.censusindia.gov.in/2011census/population_enumeration.html. Accessed 24/01/2018.

CERC (2015). Terms and Conditions for Tariff determination from Renewable Energy Sources (Fourth Amendment) Regulations, 2015. Central Electricity Regulatory Commission (CERC). http://www.cercind.gov.in/2015/regulation/SOR115.pdf. Accessed on 09.09.2018.

CII (2016). Promoting Alternative Fuel & Raw Material Usage in Indian Cement Industry. Accessed on 10th September 2018. http://www.ciiwasteexchange.org/doc/annexure_6.pdf

CPCB (2015). Central Pollution Control Board (Ministry of Environment, Forests & Climate Change). Co-processing of waste in cement plants: Status and Options to promote co-processing for waste management. Presented at Wastech 2014. Accessed on 12th September 2018. https://gpcb.gujarat.gov.in/uploads/B_Sengupta.pdf

CPCB (2016). Central Pollution Control Board (Ministry of Environment, Forests and Climate Change), Government of India, Consolidated Annual Review Report on Implementation of Municipal Solid Wastes (Management and Handling) Rules, 2000. Annual Review Report: 2014–15. April 2016. http://cpcb.nic.in/MSW_AnnualReport_2014-15.pdf. Accessed 06/12/2016.

CPCB (2017). Central Pollution Control Board (Ministry of Environment, Forests & Climate Change), Draft Guidelines for Pre-processing and Co-processing of Hazardous and Other Wastes in Cement Plants as per H&OW(M&TBM) Rules 2016. Central Pollution Control Board. Delhi, India.

CPCB (2018). Central Pollution Control Board CPCB (Ministry of Environment, Forests & Climate Change), Consolidated annual report (For the year 2016–2017) on implementation of Solid Waste Management rules, 2016 (As per provision 24(4) of SWM Rules, 2016), Parivesh Bhawan, East Arjun Nagar, Delhi - 110 032. April, 2018.

CPHEEO (2016). Municipal Solid Waste Management Manual. Part II: The Manual. Swachh Bharat Mission, Central Public Health and Environmental Engineering Organisation (CPHEEO), Ministry of Urban Development, Government of India.

CPHEEO (2018). Expert Committee Constituted by Ministry of Housing and Urban Affairs (MoHUA). Guidelines on Usage of Refuse Derived Fuel in Various Industries.

Dembla, A.K. (2017). Usage of Alternative Fuels & Raw Materials in Cement Industry as Sustainable Waste Management Option. 3rd International Conference on Alternate Fuels & Raw Materials in Cement Industry, New Delhi. http://www.cmaindia.org/upload/file/Session4-2%20Mr%20AK%20Dembla-%20Humboldt%20Wedag.pdf. Accessed 09/08/2018.

Dube, R., Nandan, V., Bineesha, P. and Dua, S. (2013). Status Paper on Utilisation of Refuse Derived Fuel (RDF) in India. GIZ GmbH under the Indo-German Environment Partnership – IGEP, New Delhi, India.

EC (2003). European Commission – Directorate General Environment, final report on Refuse Derived Fuel, Current Practice and Perspectives (B4–3040/2000/306517/MAR/E3), July 2003. http://ec.europa.eu/environment/waste/studies/pdf/rdf.pdf. Accessed 01/08/2018.

EC (2010). European Commission – Directive 2010/75/EU of the European Parliament and of the Council of 24 November 2010 on industrial emissions (integrated pollution prevention and control) https://eur-lex.europa.eu/legal-content/EN/TXT/PDF/?uri=CELEX:32010L0075&from=EN. Accessed 09/08/2018.

Flamme, S., Hanewinkel, J., Quicker, P. and Weber, K. (2018). Energieerzeugung aus Abfällen - Stand und Potenziale in Deutschland bis 2030. Umweltbundesamt (ed.) Projektnummer 75778, UBA-FB EF001021. Dessau-Roßlau, Germany.

Gallardo, A., Carlos, M., Bovea, M.D., Colomer, F.J. and Albarrán, F. (2014). Analysis of refuse-derived fuel from the municipal solid waste reject fraction and its compliance with quality standards. *Journal of Cleaner Production*, **83:** 118–125. https://doi.org/10.1016/j.jclepro.2014.07.085

GECF (2016). Environmental and Energy Policies and CO_2 Emissions. Gas Exporting Countries Forum (GECF), West Bay, Doha.

GIZ-IGEP (2013). Status Paper on Utilisation of Refuse Derived Fuel (RDF) in India. Deutsche Gesellschaft für Internationale Zusammenarbeit (GIZ) GmbH, Indo German Environment Partnership (IGEP). http://re.urban-industrial.in/content/e64918/content/e64918/live/hrdpmp/hrdpmaster/igep/content/e54413/e54441/e61172/e61173/20130425_RDFstatuspaper_final4.pdf. Accessed 09/08/2018.

H&OW (M&TBM) (2016). Hazardous and Other Wastes (Management and Transboundary Movement) Rules, 2016; Ministry of Environment, Forests and Climate Change, Government of India; G.S.R. No. 395 (E) [04-04-2016].

Hase, T., Uddin, M.A., Kato, Y., Fukui, M. and Kanao, Y. (2014). Chlorine Removal Mechanism from Municipal Solid Waste Using Steam with Various Temperatures. Energy Fuels 28:6475–6480. https://doi.org/10.1021/ef5011379

Hiebel, M., Nühlen, J. and Bertling, J. (2017). Zur Rolle der thermischen Abfallbehandlung in der Circular Economy. Fraunhofer-Institut für Umwelt-, Sicherheits- und Energietechnik UMSICHT (ed.), Kurzstudie im Auftrag der AGR Abfallentsorgungs-Gesellschaft Ruhrgebiet mbH. Oberhausen, Germany.

Hoornweg, D. and Bhada-Tata, P. (2012) What a waste – A Global Review of Solid Waste Management. Urban Development Series. March 2015, No. 15. World Bank. Washington, USA.

Iacovidou, E., Hahladakis, J., Deans, I., Velis, C. and Purnell, P. (2018). Technical properties of biomass and solid recovered fuel (SRF) co-fired with coal: Impact on multi-dimensional resource recovery value. *Waste Management*, **73:** 535–545. https://doi.org/10.1016/j.wasman.2017.07.001

IFC (2017). Increasing the Use of Alternative Fuels at Cement Plants: International Best Practice. International Finance Corporation (IFC), World Bank Group. http://documents.worldbank.org/curated/en/563771502949993280/pdf/118737-REVISED-Alternative-Fuels-08-04.pdf. Accessed 31/07/2018.

Joshi, R. and Ahmed, S. (2016). Status and challenges of municipal solid waste management in India: A review. *Cogent Environmental Science*, **2(1):** 1139434. https://doi.org/10.1080/23311843.2016.1139434

Indian Solid Waste Management Rules (SWM). Notification. Ministry of Environment, Forest and Climate Change. New Delhi, the 8th April, 2016. The Gazette of India. Regd. No. D. L.-33004/99.

Kannan, V. (2017). Waste Mapping and Forecasting for Alternate Fuel Usage in Cement Plants. Promoting Alternate Fuel and Raw material (AFR) utilisation in Indian Cement Industry. 3rd International Conference on Alternate Fuels & Raw Materials in Cement Industry, New Delhi. http://www.cmaindia.org/upload/file/Session3-3%20Mr%20V%20Kannan%20CII%20-%20Presentation%20CMA%20Conference,%20New%20Delhi.pdf. Accessed 09/08/2018

Kara, M., Günay, E., Tabak, Y. and Yıldız, S. (2009). Perspectives for pilot scale study of RDF in Istanbul, Turkey. *Waste Management*, **29:** 2976–2982. https://doi.org/10.1016/j.wasman.2009.07.014

Ketelsen, K. (2011). Stand der Restabfallvergärung in Deutschland, 8. Biogastagung, 28.-29.09.2011, Dresden.

Ketelsen, K. (2012). Beitrag der MBA zum Klima-und Ressourcenschutz, 9. Internationale ASA-Recyclingtage, 29.2.-02.03.2012, Hannover.

Ketelsen, K. (2013). Stand und Entwicklungsszenarien für die MBA in Deutschland, 25. Kasseler Abfall- und Bioenergieforum - 2013 (März 2013).

Ketelsen, K. (2015). Status and new trends / perspectives of MBT in Germany. In: Kühle-Weidemeier, Balhar (Hrsg.): Waste-to-Resources 2015, 6. International Conference, 05.-07.05.2015, Cuvillier Verlag Göttingen.

Ketelsen, K., Kannng, K., Cuhls, C. (2011). Optimisation of MBT considering energy efficiency and protection of resources and climate, International conference on waste-to-resources, 24.-27.05.2011, Hannover.

Kumar, S., Smith, S.R., Fowler, G., Velis, C., Kumar, S.J. Arya, S., Rena, Kumar, R. and Cheeseman, Ch. (2017). Challenges and opportunities associated with waste management in India. Royal Society Open Science, Published 22 March 2017. DOI: https://doi.org/10.1098/rsos.160764

LAGA (2019). Länderarbeitsgemeinschaft Abfall *LAGA PN 98. Richtlinie für das Vorgehen bei physikalischen, chemischen und biologischen Untersuchungen im Zusammenhang mit der Verwertung/Beseitigung von Abfällen, (Directive for the procedure of physical, chemical and biological evaluations with respect of waste treatment/disposal)*, Mitteilung der Länderarbeitsgemeinschaft Abfall (LAGA) 32. https://www.laga-online.de/documents/m-32_pn98_red-aend_2019_mai_1562758999.pdf

Lingk, H. (2015). Energy Efficiency in MBS Plants/ Energiemanagementsysteme in MBAs. In: Kühle-Weidemeier, Balhar (Hrsg.): Waste-to-Resources 2015, 6. Internationale Tagung MBA, Sortierung und Recycling, 05.-07.05.2015, Cuvillier Verlag Göttingen.

Ma, W., Hoffmann, G., Schirmer, M., Chen, G. and Rotter, V.S. (2010) Chlorine characterization and thermal behavior in MSW and RDF. Journal of Hazardous Materials 178(1–3):489–498. https://doi.org/10.1016/j.jhazmat.2010.01

Ma, W., Rotter, S., Hoffmann, G., Lehmann, A. (2008) Origins of chlorine in MSW and RDF: Species and analytical methods. Waste Management and the Environment IV 109:551–558. https://doi.org/10.2495/WM080561

Mani, S. and Singh, S. (2016). Sustainable municipal solid waste management in india: A policy agenda. *Procedia Environ Sci*, **35:** 150–157. https://doi.org/10.1016/j.proenv.2016.07.064

McDougall, F., White, P., Franke, M. and Hindle, P. (2003). Integrated Solid Waste Management: A life cycle inventory. 2nd edition. Blackwell Publishing Company, Oxford, UK.

Michaelowa, A., Feige, S., Honegger, M., Henzler, M., Janssen, J., Kabisch, S., Sanghal, A., Sharma, S., Pravinjith, K.P. and Kumari, A. (2015). Feasibility Study for a Waste NAMA in India. Berlin, Germany. https://www.adelphi.de/de/system/files/mediathek/bilder/Feasibility%

20Study%20for%20a%20Waste%20NAMA%20in%20India%20-%20adelphi-Perspectives.pdf. Accessed 09/08/2018.

MNRE (2018). Programme on Energy from Urban, Industrial and Agricultural wastes, Circular F. No. 20/222/2016-17-WTE. https://mnre.gov.in/sites/default/files/schemes/programme_energy-urban-industrial-agriculture-wastes-2017-2020_0.pdf. Accessed on 25.07.2018.

MoEFCC (2016). Indian Solid Waste Management Rules (SWM). Ministry of Environment, Forest and Climate Change (MoEFCC), Notification. New Delhi, 8th April, 2016. The Gazette of India. REGD. NO. D. L.-33004/99.

MSWM Rules (2000). Municipal Solid Wastes (Management & Handling) Rules, 2000. Central Pollution Control Board (Ministry of Environment & Forests) Parivesh Bhawan, East Arjun Nagar, Delhi-110 032. http://www.kspcb.gov.in/CORE%20COMMITEE-MSW.pdf. Accessed on 28.08.2018.

Parlikar, U. (2017). Swatchh Bharat Mission Management of SCF through Cement Kiln Co-processing - Issues and Way Forward. Geocycle India. 3rd International Conference on Alternate Fuels & Raw Materials in Cement Industry, New Delhi. http://www.cmaindia.org/upload/file/Session5-2%20Mr%20Ulhas.pdf. Accessed 09/08/2018

PIB (2016). Press Information Bureau, Ministry of Environment and Forests and Government of India. Solid waste management rules revised after 16 years; Rules now Extend to Urban and Industrial Areas, Javadekar. http://pib.nic.in/newsite/PrintRelease.aspx?relid=138591. Accessed 06/01/2018

Planning Commission (2014). Report of the Task Force on Waste to Energy (Volume I, II & III), Constitutions of tasks force on waste to energy projects, PPP & Infrastructure Division, Planning commission, New Delhi, India.

Rada, E.C. and Andreottola, G. (2012). RDF/SRF: Which perspective for its future in the EU. Editorial. *Waste Management*, **32:** 1059–1060. https://doi.org/10.1016/j.wasman.2012.02.017

Ragazzi, M. and Rada, E.C. (2012). RDF/SRF evolution and MSW bio-drying. In: Popov, V., Itoh, H., Brebbia, C.A. (eds.) Waste Management and the Environment VI. WIT Press. Southampton, UK. ISBN: 978-1-84564-606-6.

Rotter, S. (2010). Incineration: RDF and SRF – Solid Fuels from Waste. *In:* Christensen, T. (ed.) Solid Waste Technology & Management, 1 & 2. https://doi.org/10.1002/9780470666883.ch32

Sabery, F. (2004). Modell zur Vorkalkulation von mechanisch-biologischen Restabfallbehandlungsanlagen zur Herstellung von Ersatzbrennstoffen, PhD Dissertation, Von der Fakultät III-Prozesswissenschaften der Technischen Universität Berlin, Germany. https://pdfs.semanticscholar.org/6c9e/cb76e9ee7163dfdc4468bfe61c5fe441260b.pdf?_ga=2.139805680.218036569.1577573505-1543551893.1575553389

Sarc, R. and Lorber, K.E. (2013). Production, quality and quality assurance of Refuse Derived Fuels (RDFs). Waste Management 33:1825–1834. https://doi.org/10.1016/j.wasman.2013.05.004

SFC Environmental Technologies Pvt. Ltd. (2017a). RDF generation from India's First MSW Treatment Facility. 3rd International Conference on Alternate Fuels & Raw Materials in Cement Industry, New Delhi. http://www.cmaindia.org/upload/file/Session5-4%20Ms%20Aparna-Goa%20MSW%20RDF.pdf. Accessed 08/08/2018.

SFC Environmental Technologies Pvt. Ltd. (2017b). Technology for Municipal Solid Waste Treatment, Organic extrusion segregation technology, Mumbai, India. https://sfcenvironment.com/. Accessed 15/08/2018.

Sharma, N.K. (2017). Co-processing of Alternative fuels in Cement Kiln - JK Lakshmi Cement Experience. 3rd International Conference on Alternate Fuels & Raw Materials in Cement Industry, New Delhi. http://www.cmaindia.org/upload/file/Session1-2%20Mr%20Naveen%20Sharma%20-%20%20JKLCL.pdf. Accessed 08/08/2018.

Skutan, S. and Brunner, P.H. (2012). Metals in RDF and other high calorific value fractions from mechanical treatment of MSW: Analysis and sampling errors. *Waste Management and Research*, **30(7):** 645–655. https://doi.org/10.1177/0734242X12442740

Speier, C.J., Mondal, M.M. and Weichgrebe, D. (2018a). Evaluation of compositional characteristics of organic waste shares in municipal solid waste in fast-growing metropolitan cities of India. *Journal of Material Cycles and Waste Management*, **20(4):** 2150–2162. https://doi.org/10.1007/s10163-018-0757-y

Speier, C.J., Mondal, M.M. and Weichgrebe, D. (2018b). Data reliability of solid waste analysis in Asia's newly industrialized countries. *Int J Environ Waste Manag*, **22(1–4):** 124–146. https://doi.org/10.1504/IJEWM.2018.10014700

Swami, K. (2017). Climate Mitigation Actions by Co-processing of MSW based RDF in Cement Plants. Deutsche Gesellschaft für Internationale Zusammenarbeit (GIZ) GmbH. 3rd International Conference on Alternate Fuels & Raw Materials in Cement Industry, New Delhi. http://www.cmaindia.org/upload/file/Session5-3%20Mskamna%20swami-GIZ.pdf. Accessed 09/08/2018

SWM (2016). Solid Waste Management Rules, 2016, Government of India in the Ministry of Environment, Forest and Climate Change, S.O. 1357(E) [08–04–2016]

The World Bank (2015). Population Growth (Annual %), Year 2015. http://data.worldbank.org/indicator/SP.POP.GROW?end=2015&locations=IN&start=1960&view=chart&year_high_desc=false. Accessed 05/01/2018

The World Bank (2018). Country Data – India. https://data.worldbank.org/country/india. Accessed on 29.08.2018.

UBA (2018). Primärenergieverbrauch – Entwicklung und Ziele. Umweltbundesamt (UBA) https://www.umweltbundesamt.de/daten/energie/primaerenergieverbrauch#textpart-1. Accessed 15/08/2018

VDZ (ed.) (2016a). Umweltdaten der deutschen Zementindustrie 2015. Verein deutscher Zementwerke e.V. (VDZ). Düsseldorf, Germany. https://www.vdz-online.de/fileadmin/gruppen/vdz/3LiteraturRecherche/Umweltdaten/VDZ-Umweltdaten_2015.pdf. Accessed 15/08/2018

VDZ (ed.) (2016b). Zementindustrie im Überblick 2016/2017. Verein deutscher Zementwerke e.V. (VDZ). Berlin, Germany. https://www.vdz-online.de/fileadmin/gruppen/vdz/3LiteraturRecherche/Zementindustie_im_Ueberblick/VDZ_Zementindustrie_im_Ueberblick_2016_2017.pdf. Accessed 15/08/2018

Velis, C.A., Longhurst, P.J., Drew, G.H., Smith, R. and Pollard, J.T. (2010). Production and quality assurance of solid recovered fuels using mechanical-biological treatment (MBT) of waste: A comprehensive assessment. *Critical Reviews in Environmental Science and Technology*, **40 (12):** 979–1105. https://doi.org/10.1080/10643380802586980

Vodegel, S. and Davidovic, M. (2017). Differenzierung der energetischen Verwertung am Kriterium der Energieeffizienz. CUTEC Institut Clausthaler Umwelttechnik-Institut GmbH. Projektnummer: 60 3584 für: bvse - Bundesverband Sekundärrohstoffe und Entsorgung e.V., Bonn, Germany. http://www.bvse.de/images/pdf/altholz/2016_Cutec_Studie_Hochwertige_energetische_Verwertung_final_2.pdf. Accessed on 10/08/2018.

Weichgrebe, D. (2015). Kompendium Biogas, Veröffentlichungen des Institutes für Siedlungswasserwirtschaft und Abfalltechnik der Leibniz Universität Hannover (Heft 155), (2015) ISBN: 978-3-921421-85-7

Weichgrebe, D., Speier, C. and Mondal, M. (2017). Scientific Approach for Municipal Solid Waste Characterization. *In:* Goel, S. (ed.). Advances in solid and hazardous waste management, 1st edn. Capital Publishing Company, New Delhi.

Weichgrebe, D., Mondal, M., Vidyaranya, V., Speier, C., Zacharias, M., Murali, S., Peter, N., Urs, V., Bhaskar, U., Joseph, S., Mundkur, V. and Sivaram, R. (2016). Municipal Solid Waste (MSW) Management Study for West-Zone Bangalore, India. Technical Report (unpublished).

WRAP (2012). A Classification Scheme to Define the Quality of Waste Derived Fuels – October 2012. Waste & Resources Action Programme, UK. http://www.wrap.org.uk/sites/files/wrap/WDF_Classification_6P%20pdf.pdf. Accessed 31/07/2018

Chapter 5
Impact of Sandstone Quarrying on the Health of Quarry Workers and Local Residents: A Case Study of Keru, Jodhpur, India

Abhishek Singhal and Sudha Goel

1 Introduction

Stone has been a very popular building material in India for hundreds of years due to its availability in huge quantities and its remarkable quality. India produces more than 27% of the total stone produced throughout the world. Rajasthan is the largest producer of sandstone in India and sandstone produced in Rajasthan is considered to be of the highest international standards. Jodhpur is the second largest city in Rajasthan and one of the major sandstone quarrying and craftsmanship centres in India. Due to the high strength, abundance and availability of a wide variety of colours and textures, Jodhpur's sandstone is in huge demand in the domestic and international market. Due to the surge in exports in the last two decades, there has been an increase in the number of sandstone quarrying and processing units in Jodhpur especially in the areas around Fedusar, Keru and Mandore (Bhadra et al. 2007). Currently, sandstone quarrying and processing is one of the most important economic activities in Jodhpur providing jobs to millions of people in the city. About 2.5 million workers are employed in the mining sector in Rajasthan and involved in activities like drilling, stone crushing, blasting and loading-unloading of stone slab from vehicles (Mine Labour Protection Campaign 2005; Sishodiya et al. 2011).

The dark side of this stone industry is the unsustainable practices used in quarrying and the huge amounts of waste generated every year. These wastes are dumped unsafely, eventually affecting the local environment and health of the quarry workers and nearby local residents. During extraction and processing, around 50–70% of the extracted stone is wasted in the form of waste slurry and scrap stone (Papantonopoulos et al. 2007). Slurry is basically stone dust mixed with water, which is generated from the wet cutting of stone. Slurry and stone dust have very small sizes which can lead to air pollution due to the presence of particulate matter.

A. Singhal (✉) · S. Goel
Civil Engineering Department, Indian Institute of Technology Kharagpur, Kharagpur, West Bengal, India

D. Sengupta et al. (eds.), *Treatment and Disposal of Solid and Hazardous Wastes*,
https://doi.org/10.1007/978-3-030-29643-8_5

Further, the slurry dries quickly due to high temperatures in the Jodhpur region and the dust becomes airborne due to the slightest disturbance causing respiratory, ocular, or dermal irritation in the employees involved in quarrying activities. It can be a visual or respiratory burden for the local communities. Stone dust or slurry contains mostly silicate particles, many of these are less than 10 micron in size and enter the lungs directly through respiration. This leads to obstruction and restriction in respiration of the quarry workers and considerable reduction in lung capacity (Ghotkar et al. 1995; Singh et al. 2006). Silicosis and tuberculosis can potentially afflict employees involved in quarrying and transportation processes because the sandstone dust contains high percentage of silica (50–90%) in it. Long-term exposure to stone dust leads to silicosis and pneumoconiosis which are associated with breathlessness, chronic bronchitis, recurrent chest illness and heart failure (Scott and Grayson 2003). Other complications include tuberculosis, microbial infections, and chronic silicosis. Most of the workers are addicted to various forms of tobacco or drugs, which degrade their health even more. Due to excessive use of tobacco, many oral and dental health problems are also observed among these quarry workers (Solanki et al. 2014).

Due to the poor working environment and lack of use of any safety equipment, accidents occur resulting in heavy injury and threat to the lives of workers. These occupational injuries are one of the reasons for some major health problems that all developed, developing, and underdeveloped nations are currently facing (Solanki et al. 2014). Due to poverty, lack of awareness and education among the workers, the majority of the workforce is deprived of occupational health services. Every year all over the world around 100 million occupational injuries occur (Leigh et al. 1991). The hazardous conditions at the working site greatly affect the health and life expectancy of quarry workers due to long-term exposure from biological, chemical, and physical agents which appear at the workplace. The average life expectancy of a mine worker is approximately 52($\pm$12) years respectively, which is 10–12 years less than the life expectancy of workers who are not involved in any kind of mining activity (Mathur 1996; Verma et al. 2002). Due to these harsh conditions, many of the workers are affected by some health problem or are seriously injured and unable to work properly. Affected workers are unable to work and earn an income during their illness or injury, thus, creating economic pressure on their family. They, often, become a socioeconomic burden on others for the remaining part of their life. There are many other health and environment-related problems associated with quarrying which needed to be addressed. In the past, many studies were done in the Jodhpur area to evaluate the impacts of sandstone quarrying on the health of quarry workers (Sishodiya et al. 2011; Mathur 1996; Yadav et al. 2011; Chopra et al. 2012; Ahmad 2015). These studies were used to identify trends in health problems among workers from 1990 to 2017. Keru is a vital region in Jodhpur providence for extracting sandstone of various colours in huge amount but the worker population in the Keru quarrying area was never studied.

The objective of this study was to evaluate living conditions and the magnitude of health problems in the workers and nearby residents in the Keru quarry area. Workers and residents were interviewed and their responses to questions regarding their health, working environment and impact of quarrying activities on their life were collected and analysed. Extensive site survey was done to identify major reasons behind these health problems and injuries. To evaluate the potential for pollution by stone slurry particles in the environment, particle size analysis of the slurry particles was also done.

2 Description of the Study Area

Jodhpur is the second largest and most populated city in Rajasthan and is also the zonal headquarters for controlling most of the mineral-related activities in western Rajasthan. The geological formations in Jodhpur district date back to the pre-Cambrian era. The main types of rocks that are found in this area are sandstone, granite, limestone, rhyolite, phyllite, and slate (Central Ground Water Board 2013). There are many sandstone quarrying areas in the north-west and western areas around Jodhpur city. The study area is situated (26°20′40″N and 72°54′27″E) near the village of Keru and Badli, North-West of Jodhpur city (17.5 and 14.3 km away from the centre of the Jodhpur city) (Fig. 5.1b in red box). The site is located near Bikaner-Barmer National Highway Road (NH-125).

The temperature in the area can vary from 28 to 48 °C during summers and 8–28 °C during winters. The solar radiations throughout the year are very high, having annual average solar radiation of 22 MJ/m^2/day which can reach upto 26.5 MJ/m^2/day in summers. The wind speed remains quite low (5–8 km/h) during winter and high (20–30 km/h) during summer. Strong winds upto 28 km/h are often observed during June and the wind speed may reach 50–80 km/h during severe dust storms. The southwest monsoon contributes more than 85% of the total annual rainfall, extends from July to September. August is the wettest month with normal monthly rainfall of 128.9 mm. Mean annual rainfall of the city is 377.65 mm (1969–2014) whereas, the India Metrological Department (IMD) Normal Annual Rainfall is 314 mm.

There are a large number of quarries near Keru and Badli from where sandstone of different colour (pinkish brown, light brown, golden brown and light maroon) is extracted in huge amounts. Many of these quarries have depths of more than 5–10 m in the quarrying area. There are 113 stone cutting units in the study area from where sandstone is reshaped into slabs, strips and bricks for further finishing or selling. On 01-Dec-2016, the total area of the quarrying site was 14.2 square km with an outer perimeter of 82.4 km. The location and Google image of the site are shown in Fig. 5.2, which is currently situated on barren and vacant land.

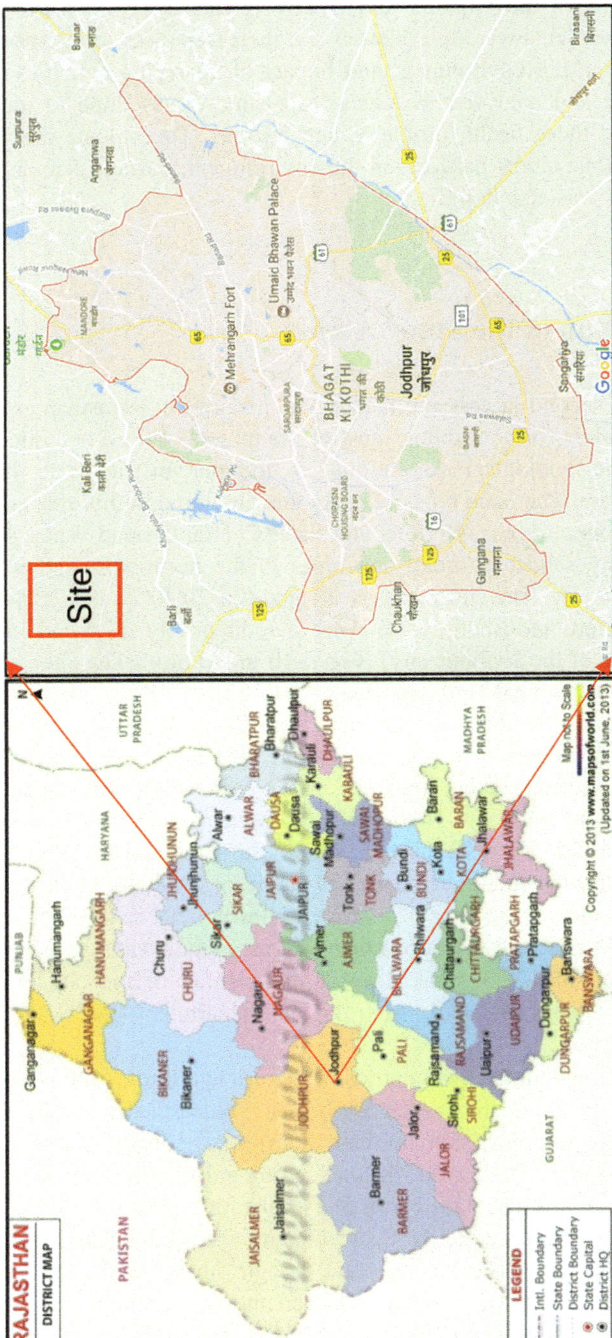

Fig. 5.1 (**a**) Location of Jodhpur city on Rajasthan state map; (**b**) Location of site with respect to Jodhpur city

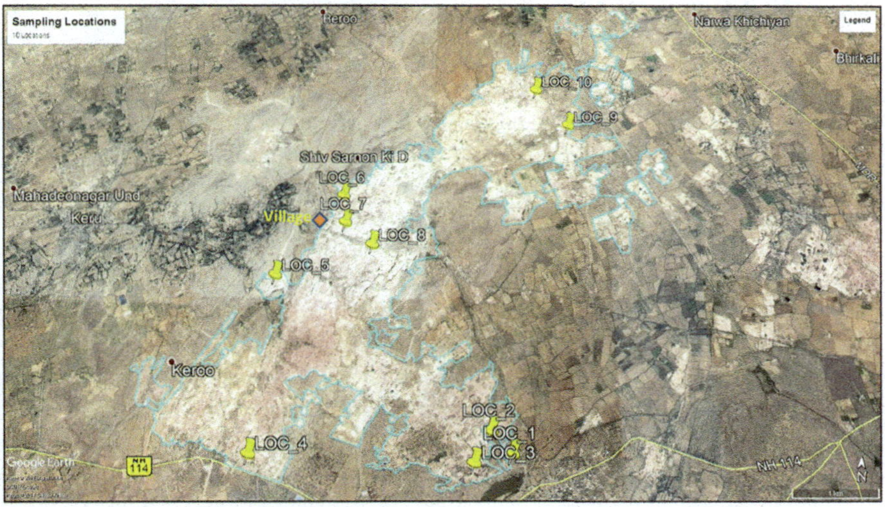

Fig. 5.2 Locations covered during survey and sample collection. (Blue line is the boundary of quarrying area)

3 Materials and Survey Methodology

3.1 Sample Collection

Samples were collected from eight different randomly selected locations in the study area as shown in Fig. 5.2. Eight slurry samples were collected from the stone cutting units for particle size analysis; two samples from LOC-1,2 and one sample each from LOC-3,4,5,7. Out of these slurry samples, five were collected directly from the cutter/cutting unit (S_4, S_6, S_9, S_{11}, S_{12}) and the remaining three (S_7, S_{10}, B_1) were collected from the slurry settling pond which is usually built near the cutting unit to separate waste slurry particles from water. All slurry samples were collected in glass bottles and preserved at low temperature.

3.2 Particle Size Analysis

All slurry samples were dried overnight and dry slurry was used for particle size analysis. Before particle size analysis, the dried slurry samples were sieved through a 4.75 mm sieve. Particle size analysis was done using Malvern Mastersizer 3000 instrument which measures particle size distributions of the sample from 10 nm to 3.5 mm by using laser diffraction technique. From the volume density curve, percentage of particles having size less than 2.5 micron and 10 micron were calculated separately to evaluate the potential for air pollution from the slurry samples in terms of PM10 and PM2.5 concentrations.

3.3 Health and Field Survey

The main aim of the survey was to assess the impact of stone waste on the health of the people working on the site, and the residents living near the quarrying area. A total of 10 quarrying and stone cutting locations and a village near the quarrying area (village near locations 6 and 7 in Fig. 5.2, village name: "Shiv sarnonkidhani") were visited for the health survey and survey of living conditions. For site survey and quarrying impact analysis, Google Earth Pro was used to obtain satellite images. Impact of waste at the site and in nearby areas was studied by visiting the quarrying site in different seasons and at different times of the day. Impact on people living nearby and working at the site was evaluated by talking to the locals. A questionnaire was prepared to evaluate the impact of stone waste and the effect of working conditions on the health of workers and residents. Specific questions were asked about health problems in workers and nearby residents, the energy sources used for cutting and drilling, the total production of stone in a day after cutting, the processes involved in mining and cutting of sandstone, the transportation system, use of safety equipment, education, and sanitation conditions at the site.

Most of the workers had problems related to breathing and eyes, so detailed questions were asked about the health problems they experience, symptoms, effect on health and frequency of medical check-up. Responses of 95 people (58 workers involved in quarrying process, 10 workers involved in stone cutting process and 27 nearby village residents) was recorded individually. Questions were also asked about working and living condition of workers and residents, impact of stone dust in different seasons, hours during which dust was less or more, noise levels and their sources, and about accidents in the past. For the study, the workers were divided into two categories on the basis of their work profile. The 1st category of workers were involved in quarrying processes and 2nd category of workers were involved in the cutting process. On the basis of their feedback, some common and specific problems related to sandstone waste and quarrying were identified. The questionnaire was prepared in English language but language of communication with the workers or residents was either Hindi or Marwari (local language).

4 Results and Discussion

4.1 Site Survey

Major activities that are part of sandstone processing and the waste generated are shown in a flowchart in Fig. 5.3.

During sandstone processing, 50–90% of the excavated sandstone is wasted in the form of scrap stone, slurry and stone dust. After excavation and cutting, sandstone is sent to the city for further processing. After quarrying and cutting, 30–50% of the excavated stone is wasted and the waste goes directly to the dumping

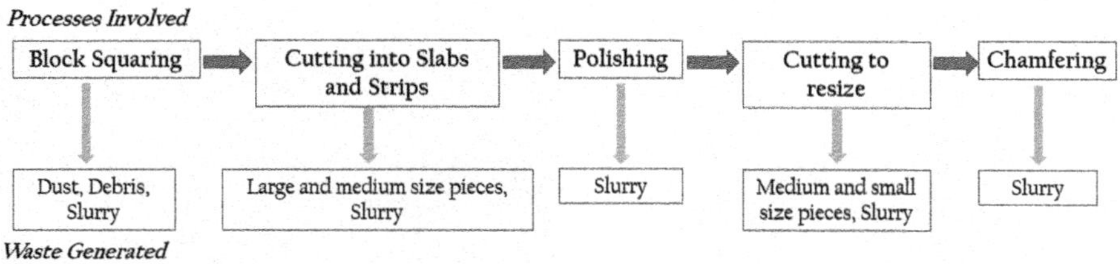

Fig. 5.3 Steps involved in sandstone processing and nature of waste generated in each step

ground near the quarrying and cutting area. Major problems related to unsustainable quarrying practices in the area are:

- Use of rapid pneumatic or electric drills (jackhammer) for stone cutting which results in huge emissions of dust in the area. Dust emissions from jackhammer and continuous movement of trucks in the area are polluting the air in this area.
- Cutting units require huge amounts of electricity, thereby increasing their carbon footprint in the environment. At many places in the study area, electricity is not available, so electricity generators are used for electricity generation which require a huge amount of diesel as fuel (6.8 L/h). These diesel-based generators run 10–12 h daily which leads to huge GHGs, PM, NOx, SOx emissions. These emissions degrade the nearby air quality which is already in poor shape due to dust emissions from various quarrying activities.
- Each cutting unit generates 8–10 MT/d of waste. There are 113 cutting units in the area which generate about 1000 MT of stone waste per day which is dumped directly in the open area.
- Trucks and tractor-trolleys are used to transport stone from quarry to cutting unit and from cutting unit to the market or for further finishing. As per local government regulations, each truck is allowed to carry no more than 10 metric tonnes of stone. However, to reduce transportation costs by reducing number of trips, owners often overload the trucks by 2–2.5 times of the allowable capacity (20–25 MT). This practice is very dangerous and has resulted in many accidents in the past due to steep slopes inside the quarrying area.
- In many areas, due to unsystematic dumping of waste in the open area, many artificial mountains of stone waste that are 5–10 m high can be seen (Fig. 5.4a). Chances of slope failure at these dumps are quite high and truck accidents occur frequently in this area.

4.2 *Particle Size Analysis*

Results of particle size analysis of slurry particles are shown in Table 5.1. The slurry had the characteristics of silt/clay and sand (slightly more silty). On an average, most slurry particles are very small, i.e., less than 330 micron (0.33 mm) in size and have very high specific surface area (94–1013 m^2/kg). Their small size indicates that these particles have high air pollution potential and their high specific surface area shows

Fig. 5.4 (a) Dumping of waste slurry and scrap stone resulting in artificial mountain of waste (5–10 m in height) (b) Dry drilling in the stone strip using a jackhammer without any safety mask or equipment

they may be high in reactivity and leaching tendency resulting in contamination of water and soil. From particle size analysis, it was found that slurry particles of size less than 2.5 microns constituted 0% to 12.74% of each sample and an average of 4.87% of the total volume of the slurry samples tested. Further, percentage of particles with size less than 10 microns ranged from 2.62% to 52.13% with an average of 20.50% of the total volume of the slurry tested. Presence of such particulate matter in slurry can lead to air pollution and health problems if the person is exposed to this material for a prolonged period of time.

In the quarries, large amounts of stone waste slurry are produced on a daily basis. Each cutting unit is used to shape around 10–20 tonnes/day of sandstone. The slurry and scrap stone from these cutting units ends up in open dumps where some of the stone dust/dry slurry is scattered by wind or rainwater and spreads over the whole quarrying area (Fig. 5.8b). From experiments and field observations, it was apparent that these slurry particles are very small and are likely to quickly disperse in the air with very little disturbance. Due to constant disturbance by wind, dry-drilling machines (jackhammer) (Fig. 5.4b) and transportation vehicles, there are always labourers exposed to airborne dust on a regular basis. During the survey, it was found that 65.52% of the workers involved in quarrying are affected by some kind of breathing problem (Fig. 5.7a). Most common health problem among workers is silicosis, which is a result of long-term exposure to stone dust which contains mostly silica (50–80% of total composition). Presence of significant amount of particulate matter in the dry slurry and stone dust and long-term exposure (8–10 h daily) may be the reason for breathing problems among quarry workers. Many dumps are situated 200–500 m away from the residential areas in the village (Fig. 5.5). Due to high wind velocity and open area, these stone dust and slurry particles also reach nearby residential areas causing pollution in the local residential areas.

Table 5.1 Results of particle size analysis of slurry samples

Parameter	B1	S6	S7	S9	S10	S11	S12	S4	Average
Gravel % (>4.75 mm)	0	0	0	0	0	0	0	0	0
Sand % (0.075–4.75)	68.60	62.43	49.29	27.32	35.94	37.29	3.41	29.12	39.175
Silt/clay % (<0.075 mm)	31.40	37.57	50.71	72.68	64.06	62.71	96.59	70.88	60.825
D10 (in mm)	0.030	0.032	0.018	0.006	0.005	0.004	0.003	0.002	0.013
D50 (in mm)	0.126	0.11	0.085	0.045	0.05	0.044	0.012	0.054	0.066
D90 (in mm)	0.525	0.239	0.214	0.161	1.06	0.19	0.046	0.212	0.330
Uniformity	1.78	0.57	0.71	1.05	5.23	1.41	1.13	1.71	1.70
Specific surface area m^2/kg	134.8	94.6	172.9	459.7	456.5	630.2	1013	422.5	423
Particles less than 2.5 μm, %	1.1	0	1.3	5.2	4.6	8.4	12.7	5.7	4.9
Particles less than 10 μm, %	4.8	2.6	6.7	20.2	21.3	30.5	52.1	25.7	20.5

Fig. 5.5 Location of dumpsite close to village houses (150–200 m away, the height of dumpsite is approximately 8–10 m, near village 'Shiv Sharnon ki Dhai')

4.3 Survey Results

Responses of workers and owners of quarrying and cutting sites about different health problems are summarized in Fig. 5.6. Responses from 58 quarry workers, 10 workers who are involved in cutting and 27 nearby village residents were collected. Workers who are involved in cutting were not affected by any health problem related to sandstone processing. Therefore, the major focus of the study was on 58 workers who were involved in the quarrying process. All labourers who participated in the survey were male and those involved in quarrying belonged to the age group of 20–50 years and workers involved in the cutting process were in the age group 20–60. Most of the quarry workers who took part in the survey were illiterate, especially those who are older than 30 years. Both groups of workers worked for 8–10 h a day (Table 5.2). Labourers involved in quarrying process work from 8:00 AM to 5:00 PM while workers involved in cutting start their work late compared to the quarry workers and work from 9:30 AM to 6:30 PM.

4.3.1 Workers and Local Residents Affected by Health Problems

Based on the responses, minimum 20% and maximum 87.5% of the labourers involved in drilling and other quarrying processes at ten different locations were affected by some breathing problem (either silicosis or tuberculosis or asthma); these results are shown in Fig. 5.6a. No health issues were found in workers involved in the cutting process. Problems related to eyes (poor vision, irritation or infection) were recorded as minimum 0% to maximum 37.5% (Fig. 5.6b). Workers affected by both health problems are a minimum of 0% to a maximum of 37.5% (Table 5.3). Based on these results, 65.5% of the workers who took part in the survey are affected by some breathing problems, 19.0% are affected by problems related to eyes and 13.8% are affected by both breathing and eye problems.

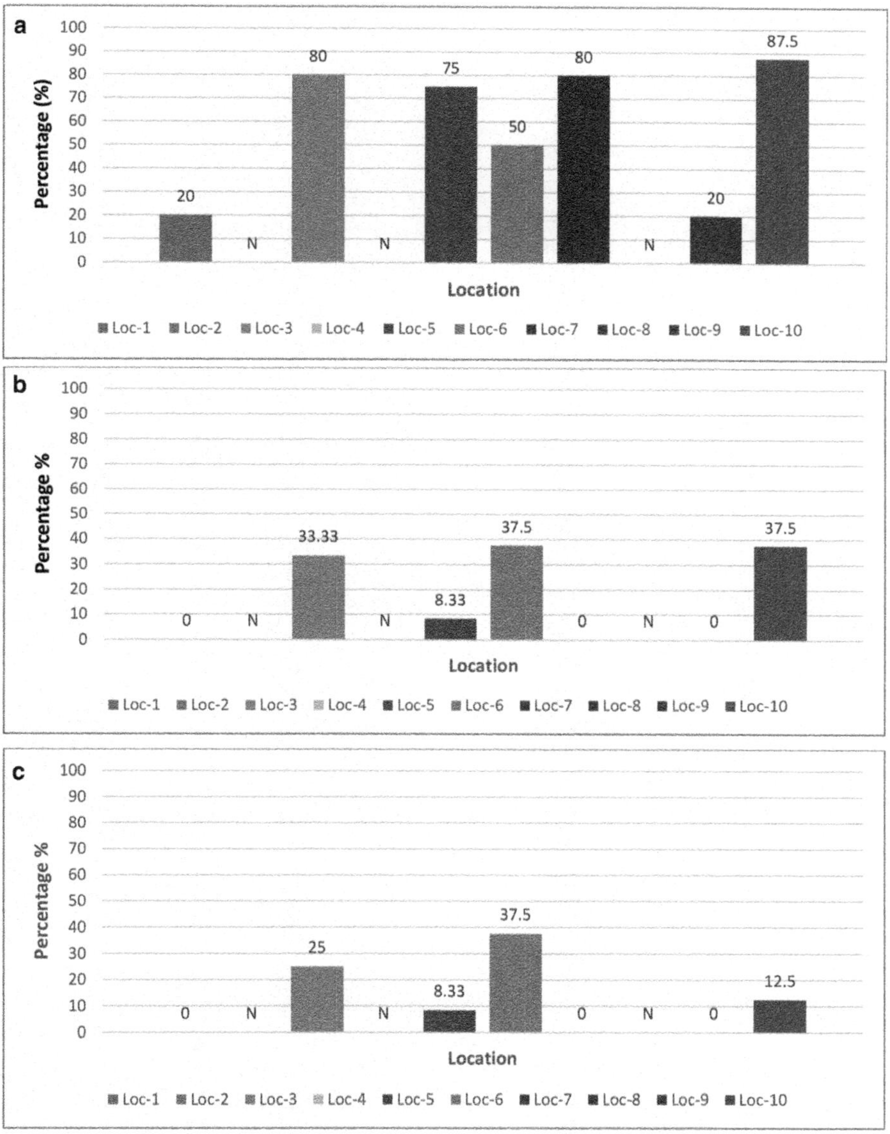

Fig. 5.6 Percentage of positive responses from quarry workers at the observed locations about (**a**) breathing problems, (**b**) problems related to eyes and (**c**) having both ("N" shows no response for this site)

A survey on quarry workers health was conducted in Jodhpur in 2015 and it was found that 70.59% of workers who took part in the survey were affected by lung disease and breathing problems (Ahmad 2015). Another study conducted on stone mine workers in Karauli district, Rajasthan shows 78.5% (73 out of 93 subjects) were affected by breathing problems of which silicosis was the most common

Table 5.2 Demographic variables of the study population

Variables	N = 58
Age	
Below 20	8 (13.79%)
20–35	23 (39.66%)
Above 35	27 (46.55%)
Gender	
Male	58 (100%)
Female	0 (0%)
Education	
Illiterate	34 (58.62%)
Primary education	23 (39.66%)
Graduate	1 (1.72%)
Tobacco or alcohol user	49 (84.48%)
Number of years of working	
1–5	16 (27.59%)
5–10	8 (13.79%)
More than 10	34 (58.62%)
Working hours	
Less than 8 h	7 (12.07%)
8–10 h	48 (82.76%)
More than 10 h	3 (5.17%)
Monthly income (in Rs)	
<5000	3 (5.17%)
5000–10,000	22 (37.93%)
>10,000	33 (56.90%)
Health problems related to	
Breathing	38 (65.5%)
Eyes	11 (19.0%)
Both	8 (13.8%)

(Sishodiya et al. 2011). Results in the current study are comparable to the results found in Ahmad (2015) and Sishodiya et al. (2011). In another study conducted by Desert Medicine Research Centre (DMC), Jodhpur in 1992–1994, 25.5% positive cases of silicosis or tuberculosis (TB) were found in the sandstone quarry workers who took part in health survey. Another study conducted by Mathur (1996) in the sandstone mines of Jodhpur, showed positive cases of TB and silicosis in 42% (120 out of 288 subjects) of sandstone workers who took part in the survey. Another study conducted by Chopra et al. (2012) showed that out of the 300 workers who participated in the study, 143 (47.7%) had sputum that was positive for silico-tuberculosis. Since 1992 to the present time, all studies show that with increment in time, the number of quarry workers with breathing problems is increasing (Table 5.4). The increase in breathing problems can be directly related to an increase in quarrying activities in the area without proper safety measures. Due to huge

Table 5.3 Responses collected from labourers at quarry site regarding health problems

Site location	Total number of response collected	Positive response on breathing problems	Positive response on vision problem	Both (breathing and visual)
Loc-1	5	1	0	0
Loc-2	N	–	–	–
Loc-3	15	12	4	3
Loc-4	N	–	–	–
Loc-5	12	9	1	1
Loc-6	8	4	3	3
Loc-7	5	4	0	0
Loc-8	N	–	–	–
Loc-9	5	1	0	0
Loc-10	8	7	3	1
Total	58	38 (65.52%)	11 (18.97%)	8 (13.79%)

Table 5.4 Summary of the literature regarding incidence of breathing-related problems in sandstone quarries

Year	Workers affected from breathing problems (in %)	Increment compared to past years available data (in %)	Increment compared to 1992–1994 (in %)	References
1992–1994	25.5	–	–	Mathur (1996)
1996	42	64.71	64.71	Malik (2005)
2012	47.67	13.5	86.94	Chopra et al. (2012)
2015	70.59	48.08	176.82	Ahmad (2015)
Current-2017	65.52	−7.018	156.94	Singhal (2018)

demand for Jodhpur sandstone in national and international markets, extraction of sandstone has increased many-fold compared to the 1990s. Since 1990, the quarrying area has increased by 4.55 times (Fig. 5.7). In the present study, there is a slight decrease in the percentage of workers affected by breathing problems as compared to a study conducted in 2014 (Ahmad 2015). This may be due to improvement in living and/or working conditions or the result of a smaller population sample size in the current study.

Probable reasons for these health problems are poor working environment, poor quality of machinery used at the site and habit of not using any safety equipment

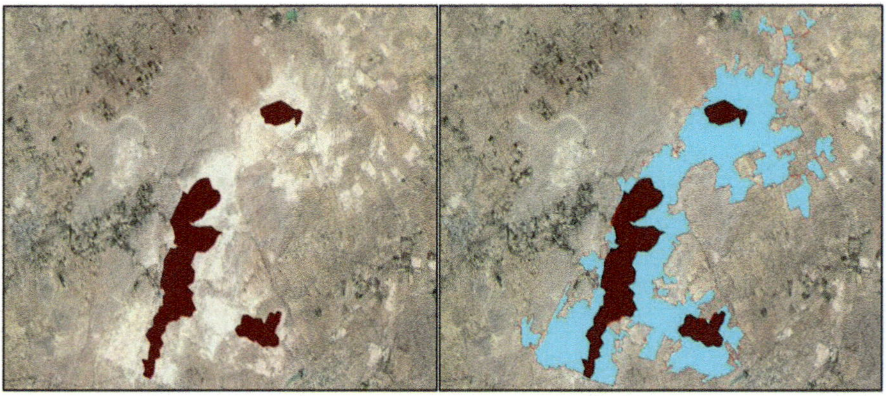

Fig. 5.7 Increment in quarrying area from 1990 to 2017 (3.12 km^2 area in 1990 and 14.2 km^2 in 2017)

during work. Rapid pneumatic or electric drills (jackhammer) are used for cutting the stone and they are run by diesel-based electricity generators. These drills generate large amounts of stone dust and most workers operate them without using any protective masks and goggles (Fig. 5.4b). These quarry workers are also exposed to stone dust at the time of loading/unloading of stone in the truck and movement of heavy vehicles releasing dust for 8–10 h every day. This long-term exposure may lead to health problems like asthma, silicosis, tuberculosis, irritation and infection in eyes. Most of the workers at the site are poor and uneducated due to which only some of them visit a doctor for diagnosis and treatment of their ailments. The rest do not get any medical attention or opt to do nothing even after being diagnosed. Common symptoms reported by workers are chronic coughing, chest pain, problems in breathing and feeling dizzy especially when working near dust. These symptoms are related to silicosis or silico-tuberculosis. Those who visit a doctor are often diagnosed with asthma, tuberculosis and silicosis (10 out of 58). Many workers have the same symptoms but they have not been diagnosed by any doctor. Therefore, there is no proof that the workers are suffering from asthma or acute silicosis or tuberculosis. Persons involved in quarrying for more than 10 years have very high chances of having silicosis or silico-tuberculosis, but cases of less than 10 years of exposure are also seen frequently (Sishodiya et al. 2011). Silicosis is a non-curable life-threatening disease and prevention is the only option for its control (Yadav et al. 2011).

The quarry workers complained about problems related to eyes during the survey. Common problems are irritation in eyes, eye infection, poor vision and swelling in eyes. The major reason behind these problems is the same as for breathing problems, which is working in a dusty environment and lack of safety equipment like protective goggles during work. None of the workers involved in drilling was observed to be using any kind of protective gear, which is necessary for protection of eyes during the drilling process. Also, at some locations, the stone is dipped in a solution of stone softening powder (lime with some other chemicals) and water. If this dry stone softening powder contacts the eyes, it can result in eye problems. Many workers who have experienced this said that it causes itching or irritation in eyes for several hours or days.

In addition, it was seen during the survey that most of the workers (84.48% of the total workers who took part in the survey) are heavily addicted to tobacco, "beedi" (a local cigarette) or liquors or some kind of drug/weed (especially opium), which further worsens their health conditions. Many oral health problems are reported due to unhygienic lifestyle and excessive use of tobacco (Solanki et al. 2014). Many studies show that the incidence of silicosis, tuberculosis, silico-tuberculosis in mine workers is much higher in those addicted to tobacco, liquor, cigarettes or opium than in those who do not use these substances (Sishodiya et al. 2011; Yadav et al. 2011; Chopra et al. 2012; Mathur 2005). Reasons for excessive use of liquor or harmful drugs was that "it helps them to release tiredness from a whole day's work and helps them to sleep well at night". According to Ahmad (2015), daily expenditure on these substances is Rs 50 and results of another study in Chhattisgarh found that one third of the income of the workers was spent on alcohol (Dabhadker et al. 2013).

No health issues were reported during the survey at the cutting site, which is somewhat cleaner and less dusty compared to the quarrying area because of wet-cutting of sandstone. Also, most workers at the cutting units are relatively more educated and economically stronger compared to workers at the quarry; hence they are more attentive to their personal health and care.

Out of the 27 residents whose responses were recorded, five of them were affected by acute and major asthma. All five were in the age range of 50 and 69 years and were diagnosed by a medical doctor. They also mentioned that most of their elders (above 55 years of age) and some children (below 10 years of age) in the village were affected by asthma, common cough and chest pain (only in elders) due to the dusty situation. On the basis of these results, it can be said that elders and children are more vulnerable to respiratory problems compared to adults. None of the residents were affected by problems related to vision or eyes during the survey.

4.3.2 Problems Related to Quarrying in the Local Area

Responses of 27 residents of nearby villages were collected to find problems related to quarrying in the site area. There are three types of problems that were mentioned: health, dust and noise problems. Health effects were described in the previous section and the other two problems are described in this section.

Most people who participated in the survey mentioned that quarrying activities and movement of trucks carrying stones during the daytime releases a large amount of dust. Many sandstone waste dumps are situated in the vicinity of the village residential area (Fig. 5.8a). Villagers also reported that due to high wind velocity in the area there is a continuous input of dust into houses. This is stone dust or dry slurry having significant amounts of particulate matter contributing to air pollution.

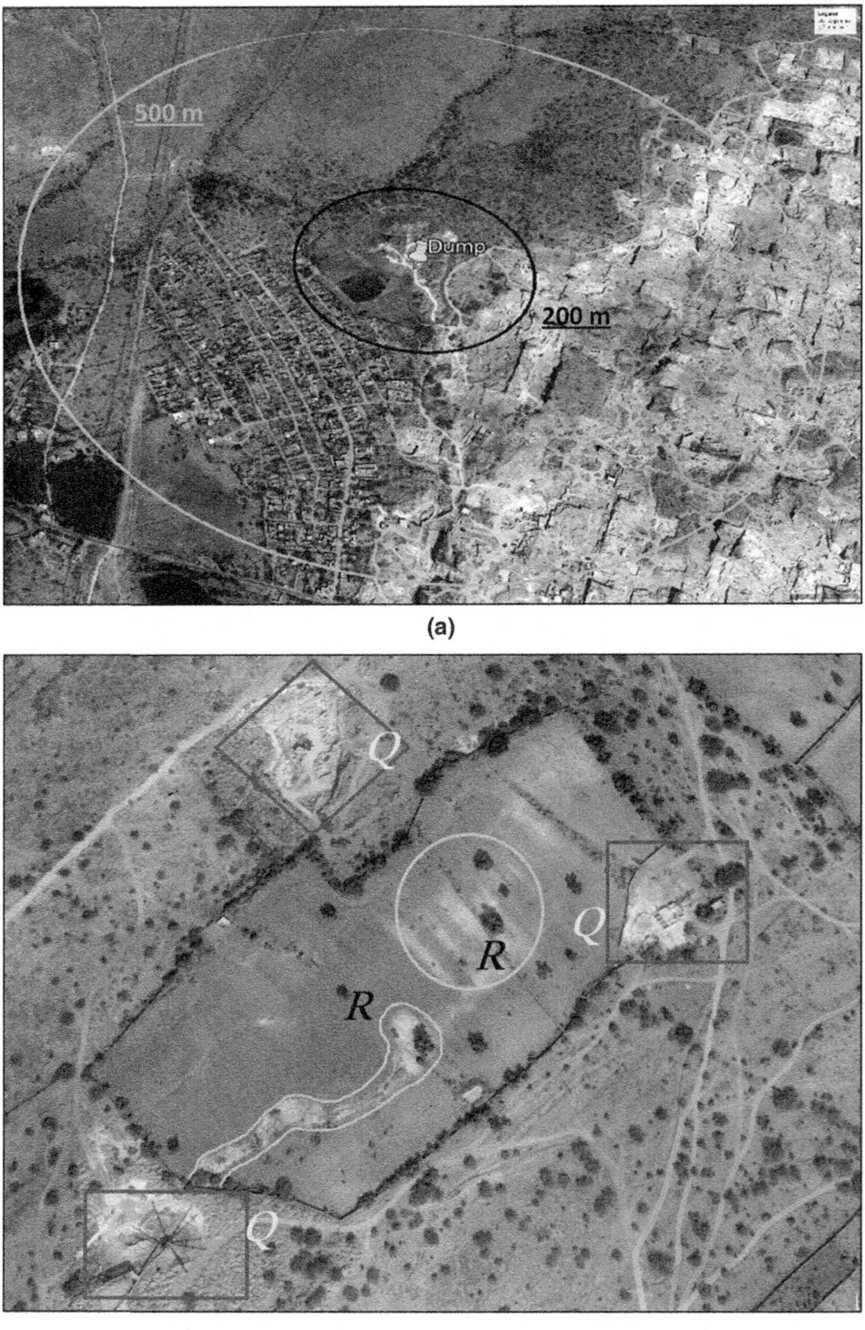

Fig. 5.8 (**a**) Satellite image showing houses and agricultural land within 200 and 500 m of the dumping area (Dump size 2.78 ha; height about 9 m). (**b**) Dust in the agricultural lands close to the quarrying areas (*Q* showing quarrying area, *R* showing stone dust in the fields due to discharge of wastewater containing stone dust/slurry and due to wind)

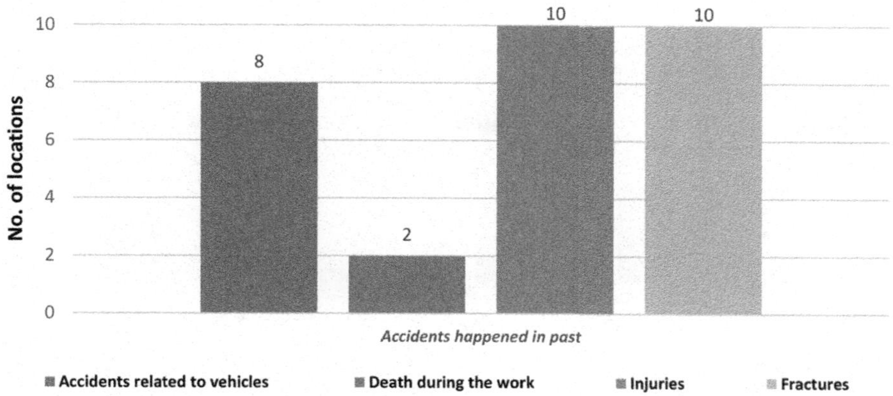

Fig. 5.9 Percentage of positive responses regarding: (**a**) seasonal variations in air-borne dust levels in area and (**b**) dust limited to working hours

The huge amounts of dust in the air eventually settle on nearby homes, vehicles and agricultural land resulting in nuisance conditions for village residents. Residents also reported that dust settles in agricultural areas resulting in poor agricultural productivity. Deposition of dust over leaves reduces stomatal conductance leading to decreased plant biomass yields and poor agricultural productivity (Zia-Khan et al. 2015).

As per resident's comments and field survey results, visible airborne dust is the main reason for respiratory problems. This dust is mainly from due to the movement of vehicles carrying stones. Responses of residents and nearby workers about daily and seasonal variations in dust levels in the air are shown in Fig. 5.9a. On being asked about the significant presence of stone dust from quarries in different seasons, all 27 people respondents said yes, the worst impact was in summers. Also, 10 people responded yes for monsoon and 20 people responded yes for winter. Also out of 27, 22 reported that conditions are dusty in the nearby area throughout the year, and are worst during sandstorms and summer seasons. Further, 18 residents responded yes to the statement that "Dust is only limited to working hours which is generally daytime and evening till 7 PM" (Fig. 5.9b).

About the problems related to noise from the quarrying process, 16 out of the 27 people responded with a yes. Normally, cutting is a noisy process but its noise cannot be heard beyond 500 m. Study of noise levels was done in various quarry areas in Jodhpur showing maximum noise level of 69.4 dBA which is less than standards (Borana et al. 2014). However, the noise from the cutter may affect the hearing capability of workers who are involved in the cutting process, yet no case of any hearing problem was reported during the survey. Trucks coming in and going out of the quarry area produce most of the noise in these villages.

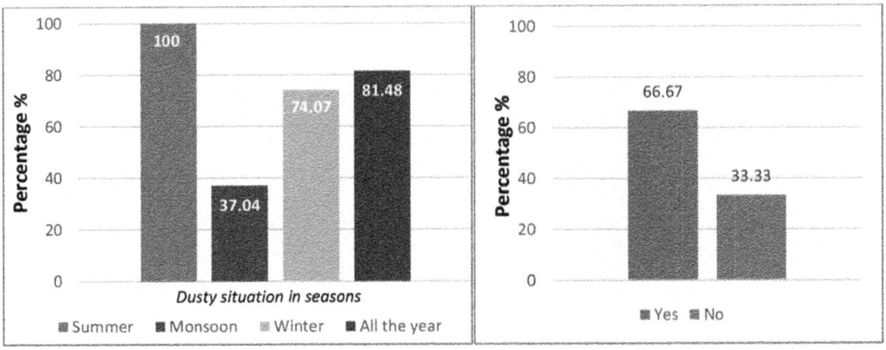

Fig. 5.10 Accidents recorded at quarrying site locations

4.3.3 Accidents in the Past Years at the Quarrying Site

Based on the survey, it can be said that working conditions at the quarry area are very unhealthy and unsafe for anyone who is working at the site. Chances of musculo-skeletal problems increase many-fold in the case of quarry workers due to these unsafe and tedious activities. Responses regarding accidents, injuries and deaths that happened in the past are summarized in Fig. 5.10.

Results of the survey show that labourers from 8 out of 10 locations had seen accidents related to vehicles in the site at some point at their life. The most common accidents reported were the toppling of trucks or tractor-trolleys during transportation or loading-unloading, stuck in a small pit during dumping waste at the dump sites, on collision with other vehicles on the highway. Overloading of trucks is also another major reason behind accidents related to driving and toppling. With overloading, chances for toppling of trucks increase many-fold on steep slopes inside the quarrying area. Some accidents were also reported at the dumping sites due to the wheel of the vehicle being stuck in the dump, which is a result of uncontrolled and unsystematic dumping of stone waste on the dump. Most of the dumping sites in the area look like artificial mountains of loose, scrap stones and stone dust which can be up to 10 m high at some locations (Fig. 5.4a). Roads built over these dumps have very loose sub-base and due to very large height and uneven slope, there is always fear of slope failure. This can lead to fatal accidents since quarry workers frequently work or rest near these dumping areas.

Results of the survey also show that people from all the sites (all 10 locations) had seen cases of bone or ligament fracture and some permanent injury in their time working at the site. Some common reasons behind these injuries are not using safety equipment during work, unsafe practices in loading-unloading of stone in vehicles, drilling process, etc. At two locations (Loc-2 and Loc-7) workers reported the deaths of some workers during work at the quarry. Main reason behind the death of these workers was slipping of the foot at the top quarry pit which is approximately 9–11 m deep from the ground level and sliding of base stone block (akin to a landslide) during drilling. Depth of the locations where quarrying activities are done are sometimes more than 10 m and quarry pits in these areas are usually filled with water which may be the reason for death or fatal injury if somebody falls into them. To prevent these fatal accidents, proper protective measures should be taken by quarry owner or government authorities.

5 Recommendations to Improve Present Conditions

On the basis of the site and health survey, it can be concluded that the current health conditions of the quarry workers and local residents is in very bad shape. Main reasons behind these problems are unsustainable quarrying practices and huge quantity of waste associated with them. First step to improve health conditions of the population is to provide proper medical care to the quarry workers and nearby residents. Also, with the help of Non-Government Organizations (NGOs) and local government authorities, educational and training programmes should be conducted regarding health impacts associated with quarrying and sandstone dust. To tackle problems associated with unsustainable quarrying practices, three measures can be taken. First and most important measure is reduction of generated waste (source reduction) and emissions by providing proper machinery and equipment. Second measure is to reuse waste or convert waste into useful construction materials. Third measure is safe disposal of waste. Suggested measures which can be taken are described here.

1. *Improvement in machinery and equipment*

   - Instead of dry-drilling or use of jackhammer, wet cutting should be used at the excavation site to reduce dust emissions. This will protect workers from dust exposure and significantly reduce the potential of deadly threats like silicosis and silico-tuberculosis.
   - Most injuries and health problems are due to lack of safety equipment. Therefore, personal protective equipment (PEE) like helmet, dust mask, goggles, and rubber boots should be provided by quarry owners and their use should be mandatory.
   - Instead of diesel generators, more power efficient battery based electricity inverters should be used for drilling. They will result in massive reduction in GHGs, PM, SO_X and NO_X emission at the working site.
   - Overloading of vehicle should be strictly prohibited and this regulation should be implemented thoroughly to prevent accidents. Also, trucks and mini-trucks should be used instead of tractor-trolleys because tractor-trolleys are not designed to carry heavy materials like stones and are unstable vehicles on loose surfaces.
   - There are several abandoned and empty quarry pits in the area which can be used for rain water harvesting. This water can be used for wet drilling and cutting, reducing dust emissions at site by water sprinkling and for other quarrying activates. Due to water scarcity in the region it could be very helpful in fulfilling water demand for quarrying.

Fig. 5.11 Reuse of scrap stone waste in boundary wall construction (**a**) with mud-water paste and (**b**) cement paste

2. Reuse of waste products

- Scrap stone and fine stone slurry can be used as construction material along with cement and concrete. Many researchers have tried to reuse marble and granite waste as aggregate in concrete (Almeida et al. 2007; Mahzuz et al. 2011). However, research on reusing sandstone waste in concrete is still limited. Aciu (2014) have tried to use fine sandstone slurry waste as 33% and 50% replacement of fine aggregates in the mortars resulting in significant improvement in compressive and bending strength with better adhesion of aggregates and cement. Sandstone dust can replace sand to a certain extent and can be used for local and government construction in the nearby areas.
- Scrap stone and fine stone dust can be used for constructing low-cost masonry wall by using stone dust as a fine aggregate and scrap stones as coarse aggregate. In some rural areas of Jodhpur district, it is used for building walls by piling scrap stones on each other and binding them with a mixture of mud and water (Fig. 5.11).
- Fine stone dust and dry slurry can be used for strengthening the base layer in road construction. According to Papantonopoulos et al. (2007), Almeida et al. (2007) and Al-Joulani (2012), using fine stone dust below the road layers reduces water intrusion in base layer and resulting in more strength in base layers. Very small size of stone dust leads to reduction in permeability and hydraulic conductivity with increment in strength and bearing capacity of the soil. Also, addition of stone dust to clays reduces their plasticity and increase silty behaviour which is quite favourable in terms of workability (Sivrikaya et al. 2014).
- Waste scrap stone can be used to make low-cost pathways in gardens or by the roadside. Different natural colour availability in sandstone provide better look and also eliminates costs associated with artificial colouring.

3. *Sustainable dumping of stone waste*

- A large number of dumping locations exist in the vicinity of the quarry areas. Dust emissions from these gigantic dump sites is a major problem due to high wind velocities in this region. So, dumping locations should be provided 500–1000 m away from the residential area or working place. The stone waste can be used to fill empty quarries in the area.

- Most of the dumping locations in the area are of gigantic size and upto 8–10 m high. This poses the threat of slope failure while dumping. So, design and slope of these dump sites should be improved. Designing a proper landfill for dumping stone waste is necessary.

6 Conclusions

The results obtained from particle size analysis, site survey and health survey show that there are serious adverse effects on the health of quarry workers, residents and on the environment at the Keru site. Result of site survey shows that fine stone slurry is the major pollutant and current unsustainable quarrying practices are the main reason behind the environmental degradation and health problems in the area. Particle size analysis shows that dry slurry has very small particles, mostly less than 330 micron (0.33 mm). Slurry contains small amounts of particles of size less than 2.5 micron (average of all tested slurry samples was 4.87%, maximum was 12.74%) and significant amounts of particles of size less than 10 micron (average of all tested slurry samples was 20.50%, maximum was 50.13%). Quarrying, transportation activities, local wind effects and the presence of significant amounts of PM10 particles in slurry/stone dust make dust a major air pollutant at this site.

Results of the health survey also show that out of the 58 quarry workers who took part in the survey, 65.52% workers were affected by some kind of breathing problem, 18.97% were affected by some kind of eyesight problem and 13.79% were affected by both problems. Survey of the site showed that poor quality of machinery used with no safety equipment has led to injuries and major health problems among quarry workers. Results of the health survey are comparable with recent studies in the area and show an increment in comparison to older survey data of the area. As per responses from nearby residents, about 18.5% of the local residents are also affected by breathing problems. Results of the survey also show that due to unhealthy and unsafe quarrying practices, there are a number of accidents in the area with the workers and vehicles resulting in fatal injuries (permanent and temporary) and even deaths in some cases. Improvement in quarrying processes and machinery with better awareness programmes regarding safety and health issues amongst workers and residents are essential for better living and health conditions.

References

Aciu, C. (2014). Research on Recycled Ceramic Waste in the Composition of Ecological Mortars. *Journal of Applied Engineering Sciences*, **4(17), 144:** pp. 7–12.

Ahmad, A. (2015). A study of Miners, Demographics and Health Status in Jodhpur District of Rajasthan, India. *Int. J. Dev. Stud. Res.*, **3:** 113–121.

Al-Joulani, N. (2012). Effect of Stone Powder and Lime on Strength Compaction and CBR Properties of Fine Soils. *Jordan Journal of Civil Engineering*, **159(697),** pp. 1–16.

Almeida, N., Branco, F. and Santos, J.R. (2007). Recycling of stone slurry in industrial activities: Application to concrete mixtures. *Building and Environment*, **42:** 810–819.

Bhadra, B.K., Gupta, A.K., Sharma, J.R. and Choudhary, B.R. (2007). Mining activity and its impact on the environment: Study from Makrana marbles and Jodhpur sandstone mining areas of Rajasthan. *J. Geol. Soc. India*, **70:** 557–570.

Borana, S.L., Yadav, S.K., Parihar, S.K. and Palria, V.S. (2014). Impact Analysis of Sandstone Mines on Environment and LU-LC Features Using Remote Sensing and GIS Technique: A Case Study of the Jodhpur City, Rajasthan, India. *J. Environ. Research Dev.*, **8:** 796–804.

Chopra, K., Prabhu, P., Bhansali, S., Mathur, A. and Gupta, P.K. (2012). Incidence & prevalence of silicotuberculosis in western Rajasthan: A retrospective study of three years. *Natl. J. Community Med.*, **3:** 161–163.

Central Ground Water Board (2013). Groundwater Scenario, Jodhpur District, Rajasthan. Jaipur: Ministry of Water Resources, Government of India.

Dabhadker, K., Shrivastava, R. and Sharma, A. (2013). Nutrition of Coal mine worker. *Int. J. Sci. Technol. Res.*, **3:** 161–163.

Ghotkar, V.B., Maldhure, B.R. and Zodpey, S.P. (1995). Involvement of lung and lung function tests in stone quarry workers. *Indian J. of Tuberc.*, **42:** 155–160.

Leigh, J., Macaskill, P., Kuosma, E. and Mandryk, J. (1991). Global burden of disease and injury due to occupational factors. *Epidemiology*, **10:** 626–631.

Mahzuz, H.M.A., Ahmed, A.A.M. and Yusuf, M.A. (2011). Use of stone powder in concrete and mortar as an alternative of sand. *African Journal of Environmental Science and Technology*, **5 (5):** 381–388.

Malik, D. (2005) [cited 2018 June 10]. India together (Internet). Silicosis – A 'Dusty' tale of Rajasthan. Available from: http://www.indiatogether.org/lungdust-environment.

Mathur, M.L. (1996). Silicosis among sandstone quarry workers of a desert district Jodhpur. *Ann. Natl. Acad. Med. Sci.*, **32:** 113–118.

Mathur, M.L. (2005). Pattern and predictors of mortality in sandstone quarry workers. *Indian J. Occup. Environ. Med.*, **9:** 80.

Mine Labour Protection Campaign (2005). The mining industry of Rajasthan. School of Desert sciences, Jodhpur (India). Available from: http://asmasiapacific.com/wp-content/uploads/2014/07/BaharDutt2005_OrganisingTheUno.pdf

Papantonopoulos, G., Taxiarchou, M., Bonito, N., Adam, K. and Christodoulou, I. (2007). A study on best available techniques for the management of stone wastes. *In:* Proceedings of the 3rd International Conference on Sustainable Development Indicators in the Mineral Industries (SDIMI 2007): 17–20 June 2007, Milos Islands, Greece.

Scott, D.F. and Grayson, R.L. (2003). Selected Health Issues in Mining. Centre for Disease Control. Available from: https://www.cdc.gov/niosh/mining/userfiles/works/pdfs/shiim.pdf

Singhal (2018) Environmental impacts of sandstone quarrying at Keru, Jodhpur, Rajasthan, M. Tech. Thesis, IIT Kharagpur, India.

Singh, S.K., Chowdhary, G.R. and Purohit, G. (2006). Assessment of impact of high particulate concentration on peak expiratory flow rate of lungs of sandstone quarry workers. *Int. J. Environ. Res. Public Health*, **3:** 355–359.

Sishodiya, P.K., Nandi, S.S. and Dhatrak, S.V. (2011). Detection of silicosis among stone mine workers from Karauli District. National Institute of Miners Health, Nagpur (India). Available from: http://aravali.org.in/themes/upload/files/276725.pdf

Sivrikaya, O., Kıyıldı, K.R. and Karaca, Z. (2014). Recycling waste from natural stone processing plants to stabilise clayey soil. *Environ Earth Sci.*, **71:** 4397–4407. DOI: https://doi.org/10.1007/s12665-013-2833-x

Solanki, J., Gupta, S. and Chand, S. (2014). Oral Health of Stone Mine Workers of Jodhpur City, Rajasthan, India. *Saf. Health Work*, **5:** 136–139.

Verma, D., Purdham, J.T. and Roels, H.A. (2002). Translating evidence about occupational conditions into strategies for prevention. *Occup. Environ. Med.*, **59:** 205–214.

Yadav, S.P., Anand, P.K. and Singh, H. (2011). Awareness and Practices about Silicosis among the Sandstone Quarry Workers in Desert Ecology of Jodhpur, Rajasthan, India. *J. Hum. Ecol.*, **33:** 191–196.

Zia-Khan, S., Spreer, W., Pengnian, Y., Zhao, X., Othmanli, H., He, X. and Müller, J. (2015). Effect of Dust Deposition on Stomatal Conductance and Leaf Temperature of Cotton in Northwest China. *Water*, **7:** 116–131; doi: https://doi.org/10.3390/w7010116

Chapter 6
Incorporation of Life Cycle Thinking in Development of Integrated Solid Waste Management Systems

V. R. Sankar Cheela and Brajesh K. Dubey

1 Life Cycle Thinking: A Brief History

Scottish economist and biologist Patrick Geddes coined the concepts of life cycle thinking during 1880, with an aim to develop an inventory for coal with a focus on energy supply. During the late 1960s and early 1970s, industries started performing LCA studies with an emphasis on resource consumption, energy analysis and emissions for the product systems. This process is termed as Resource and Environmental Profile Analysis (REPA) by the Midwest Research Institute located in the United States. This process was termed as inventory studies without impact assessment; in present scenario they are known as Life Cycle Inventory Studies. The impact assessment is not a part of the study during the period 1970–1980. Due to the lack of knowledge sharing platforms for LCA, a standardized procedure was not in practice for performing the LCA studies. Hence, the outcome of the analysis conducted for the same objective and product are contrasting and contradicting.

After one decade, the terms "life cycle analysis" and "life cycle assessment" are coined in Europe and North America with an increase in environmental consciousness. In the year 1984, Switzerland has drafted a report "Environmental report on Packaging". In the year 1990, during a workshop conducted by SETAC on "A Technical Framework for Life Cycle Assessment", the committee introduced the concept of LCA triangle. Inventory, impact analysis, and improvement analysis are the components of the LCA triangle (Fig. 6.1a). Between the period 1990 and 1993, SETAC and SETAC Europe made new developments for further standardization of the LCA process. SETAC revised the LCA triangle during the workshop held at Sesimbra, Portugal in the year 1993. The team introduced a new component termed "Goal Definition and Scoping". This component is located in the middle of the LCA

V. R. S. Cheela · B. K. Dubey (✉)
Environmental Engineering and Management Division, Civil Engineering Department, Indian Institute of Technology Kharagpur, Kharagpur, West Bengal, India
e-mail: bkdubey@civil.iitkgp.ac.in

D. Sengupta et al. (eds.), *Treatment and Disposal of Solid and Hazardous Wastes*,
https://doi.org/10.1007/978-3-030-29643-8_6

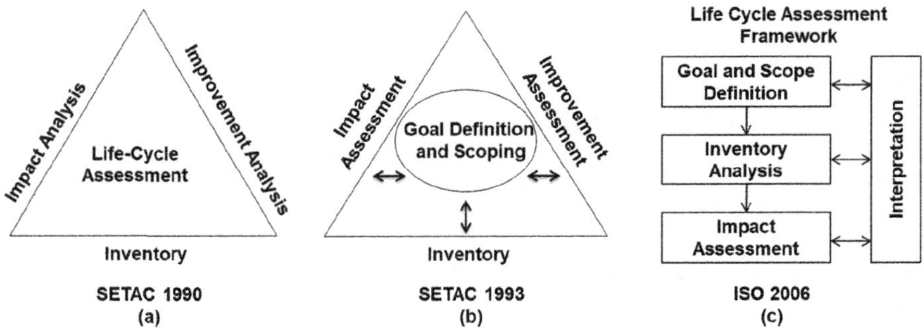

Fig. 6.1 Formulation and developments in LCA framework

triangle with interconnections to the other three elements (Fig. 6.1b). In the year 1996–1997, ISO revised the framework for the LCA (ISO 14040:2006a) and developed a flow chart representing step-by-step procedure in place of the LCA triangle (Fig. 6.1c). This framework included the direct applications, indicating that the study has to consider the intended use of results obtained. Figure 6.1 represents the formulation and developments of the LCA framework (Klopffer and Grahl 2014).

2 Life Cycle Assessment

2.1 Definition and Limitations

As per the International Standard Organization, definition of LCA is "compilation and evaluation of the inputs, outputs and potential environmental impacts of a product system throughout its life cycle" and definition of life cycle is "consecutive and interlinked stages of a product system, from raw material acquisition or generation of natural resources to the final disposal" (ISO (International Standard Organization) 1997). LCA is a systematic and scientific approach to compute, analyse, and evaluate the environmental impacts of a product based on the product itself or function provided at all stages of the product lifecycle. In a broader sense, the product includes physical goods, services and systems. The product life is a combination of the unit process involving the development, utilization, management, and disposal of the product. The unit process includes extraction of resources, raw material, production, utilization, management, and disposal of the product. The assessment of the environmental burdens includes all the impacts associated with extraction process, types of land use, raw materials, energy and emissions. The end results of the LCA approach are quantitative in character; however if a quantitative output is not possible qualitative aspects are taken into consideration for developing an overall picture in terms of environmental impacts. The final application and utilization of the product play a pivotal role in the economic growth. Cradle-to-grave approach is applied in the development of a holistic approach to determine the overall impacts from all the unit process involved in development of the product. Further this approach reduces the shifting of the impacts between the stages of product development. This is applied in strategic planning, government policies and business approaches. This includes design, development, analysis, and comparison of an existing or new product, process or a service. For example, in solid waste management, LCA study can be performed for the existing waste management system in an urban local body (ULB) (or) during the decision making process for implementing a new treatment technology (or) to compare different types of biological treatment systems (or) comparison of the waste management systems being implemented in different ULBs (Sankar and Dubey 2019).

Any approach has both sides of the coin; for LCA, its holistic nature is major strength and limitation. LCA framework will not address the localized impacts to the full extent, certain technical assumptions and choices are to be developed as transparent as possible to achieve the required results. The environmental impacts are assessed based on an arbitrarily defined functional unit and not based on spatial and temporal components. Life cycle inventory databases that are used as datasets for assessment of the potential impacts should be revised and updated over a period of time based on the advancements in technology, variations in energy production and other temporally varying factors. The format of database has to be standardized globally for performing comparative studies. The databases are developed based in the given time frame and over a period of time they become obsolete. Furthermore, development of database is time and cost extensive process. The design of the LCA model is based on the linear modelling and will not address the economic and social factors. Finally, LCA is not a decision making process by itself; it provides the information in support of decision, equipping the decision makers for better and effective planning of systems.

Risk assessment and substance flow analysis along with LCA studies provide a better understanding of the local impacts due to the core processes of individual substances. Economic studies can be conducted using Life Cycle Costing (LCC) approach and cost-benefit analysis. Social impact assessment studies for social factors can be included to develop an integrated toolbox for the overall and holistic solution. Figure 6.2 represents the ISO standardized LCA framework. The LCA study includes four stages.

Stage one is the **definition phase**. In this goal and scope are described: need, necessity and nature of the work to be performed during the study. The aim and objective of the LCA study predominantly depend on the goal. Based on the purpose of the study, subject and intended application, the scope of the LCA is determined. During this stage, system boundary, functional unit, and reference flow are determined. For comparative study, different scenarios are developed based on the combination of possible and feasible alternatives.

Stage two is known as the **inventory analysis phase**; this is the heart and core of the LCA study. In this stage, to achieve the required goal, the input and output data is acquired based on the spatial, temporal and technical constraints for each unit process. Data compilation is carried out by data collection from both primary and

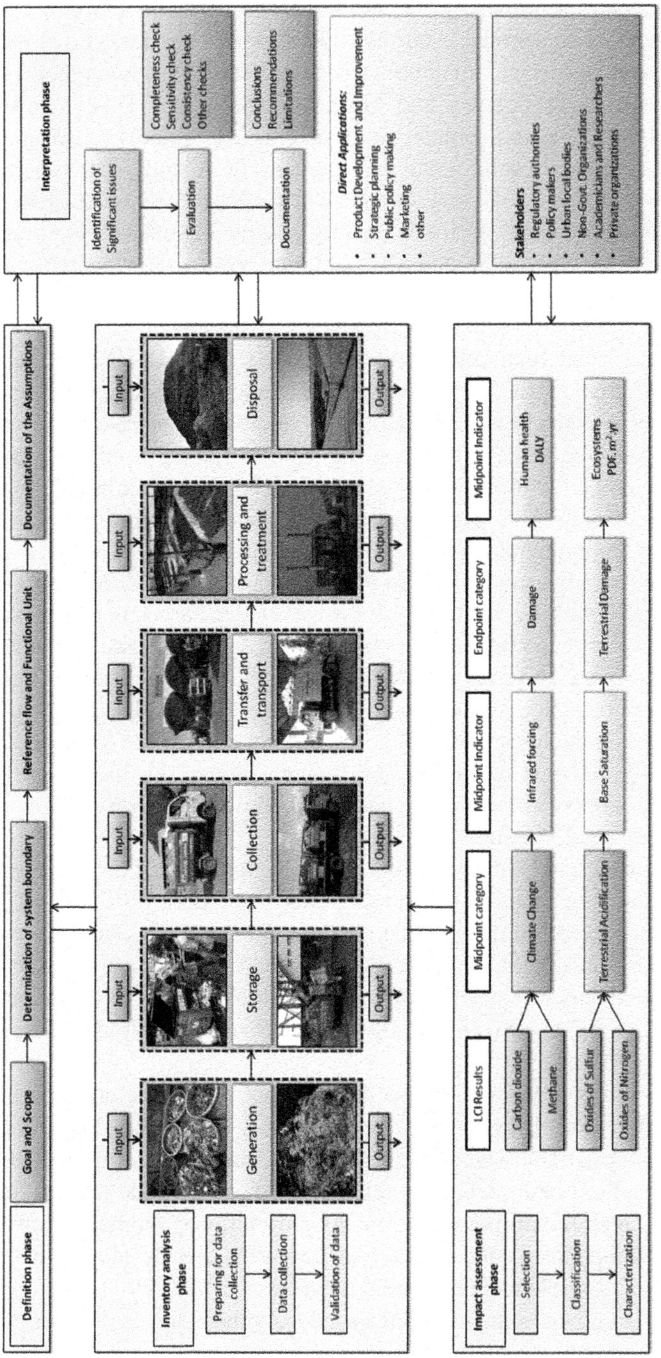

Fig. 6.2 ISO standardized LCA framework

secondary sources to develop an inventory database. This stage is also known as the Life Cycle Inventory Analysis.

Stage three is known as the **impact assessment phase**; in this stage, the assessment of environmental impacts is performed. Based on the inventory database compiled, analysis and investigations are conducted to quantify the magnitude of the environmental burdens of the product or service within the system boundary defined in stage one. This knowledge provides the basis for the designing and planning of environmentally sound systems. This stage is also known as the Life Cycle Impact Assessment (LCIA).

Stage four is known as the **interpretation phase**; this phase is a documentation phase in which the results obtained from steps two and three are discussed and summarized. Based on this, necessary conclusions and recommendations are developed as a part decision making by the goal and scope.

2.2 *Goal and Scope Definition*

In the LCA, the goal and scope is the primary component. Fundamentally, we have to identify the aim, objective and the need for performing LCA in our study. As per ISO 14040:2006a "The goal and scope of an LCA should be stated clearly by the practitioner, and it should be consistent with the intended application. Due to the iterative nature of LCA, the scope may have to be refined during the study". The stakeholder(s) commissioning or performing the LCA, defines the goal with a detailed explanation on the objective of the study (range of applications), need for the survey (interest of realization), target group(s), and accessibility to the public in the form of publication or other modes (if comparative assertions are intended). During the study, the domain knowledge gained and database created provide a better understanding of the system. At this stage, the scope can be modified by the concerned authorities based on the revised requirements. The framework standardized by ISO represents that the LCA study is iterative (double arrows). Documentation has to be done periodically to record and track the modifications in goal or scope made throughout the study.

2.2.1 Product System

Product system plays a vital role in the LCA studies. It is a grouping of unit processes, intermediate products, elementary and product flows across and within the system boundaries performing one of the more defined services. The product system is defined based on the function of the product and not by the end products. Figure 6.3 represents the product system for life cycle inventory analysis.

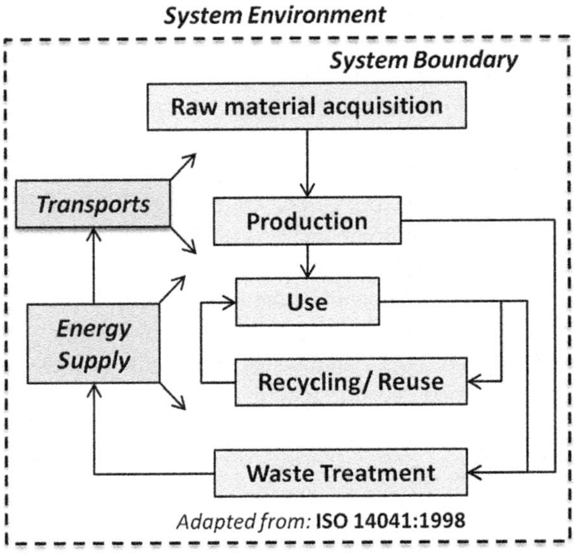

Fig. 6.3 Product systems for life cycle inventory analysis

2.2.2 System Boundary

As per ISO-14040:2006a, a system boundary is defined as "interface between a product system and the environment or other product systems". System boundary is determined based on the goal of the study. The criteria for establishing the system boundary should be justified in the scope. It describes the unit processes that are included in the LCA study. Data category, intended application, cut-off criteria, assumptions, cost, and the intended audience are the factors to determine the system boundary. The system boundary should include all the life cycle stages, processes and inputs or outputs in the development of the product. The deletion of any component from the system boundary should be substantiated with proper justification and documented. Figure 6.4 presents the types of system boundaries for LCA studies on integrated solid waste management.

The LCA study is classified based on the technical components included in the system boundary. The following sections present the description of the system boundary:

1. *Cradle-to-grave:* The system comprises the extraction of resources, manufacturing, utilization, and disposal of the product. In an SWM study, generation, collection, transfer and transport, processing, treatment and disposal of the waste are part of the system boundary. Figure 6.4 represents the system boundary for cradle-to-grave analysis.

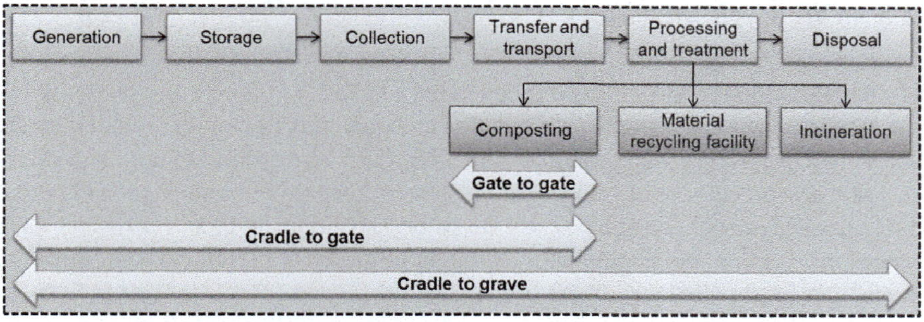

Fig. 6.4 Types of system boundaries for LCA studies on integrated solid waste management

2. *Cradle-to-gate:* The system includes extraction for resources, manufacturing, and other process involved before the product leaves the factory gate. In an SWM study, generation, collection, transfer and transport, processing, treatment and disposal of the waste are part of the system boundary. Figure 6.4 represents the system boundary for cradle-to-gate analysis.
3. *Gate-to-gate:* The system includes the manufacturing process of the product within a factory or a project site. In an SWM, if composting, bio-methanation, incineration, or material recycling facility are studied exclusively as part of this system. For example, in the composting unit for LCA studies waste entering the unit gate to the preparation of compost will be considered as the system boundary. Figure 6.4 represents the system boundary for gate-to-gate analysis.

2.2.3 Functional Unit and Reference Flow

As per ISO-14040:2006a, a functional unit is defined as "quantified performance of a product system for use as a reference unit in a life cycle assessment study". The functional unit provides scope for normalization of the input and output data. Solid waste management is one of the essential functions of an urban local body. The primary elements of the SWM system include storage, collection, transfer and transport, processing, treatment and disposal of waste. Based on the resources availability, economy and characteristics of the waste, ULBs are implementing different methods and approaches. The ULBs implement individual or combination of material recovery, composting, bio-methanation, incineration process for the treatment of the waste in the ULB. Hence, for performing a full scale LCA study or comparative LCA study, the function and the functional unit must be defined initially. In the SWM-LCA studies, management of the waste is the function. The functional unit can be one ton of garbage or the total amount of waste generated in a year. Once the functional unit is defined, it provides a basis for the determination of the input(s) and output(s) for the unit operations specified in the reference flow.

2.3 Life Cycle Inventory Analysis

As per ISO 14041:1998, life cycle inventory analysis is defined as "phase of life cycle assessment involving the compilation and quantification of inputs and outputs, for a given product system throughout its life cycle". Life cycle inventory analysis is a systematic process for developing an inventory database from the input and output flows associated with the unit process. Inventory development process involves three stages of planning, collection, and validation of the data. Figure 6.5 represents the flow chart for the life cycle inventory analysis.

In the planning stage, based on the goal and scope of the LCA study, data to be collected for the foreground and background systems is determined. Foreground data is data associated with the unit process involved in the system boundary. Raw materials, energy, water and emissions into the air, water and soil are examples for foreground data. This dataset has a direct influence on the results obtained from the impact assessment study. Background data is data associated with the process that is allied with the unit process involved in the system boundary, for example, manufacturing of storage bins, diesel production (collection) and manufacturing of trucks (transport), etc. This dataset has indirect, direct or zero influence on the impact assessment study. In a full-scale LCA study, the background data can be considered based on the goal and scope definition. While in the comparative LCA study this data is neglected as this data is listed under the common dataset for different system boundaries under consideration.

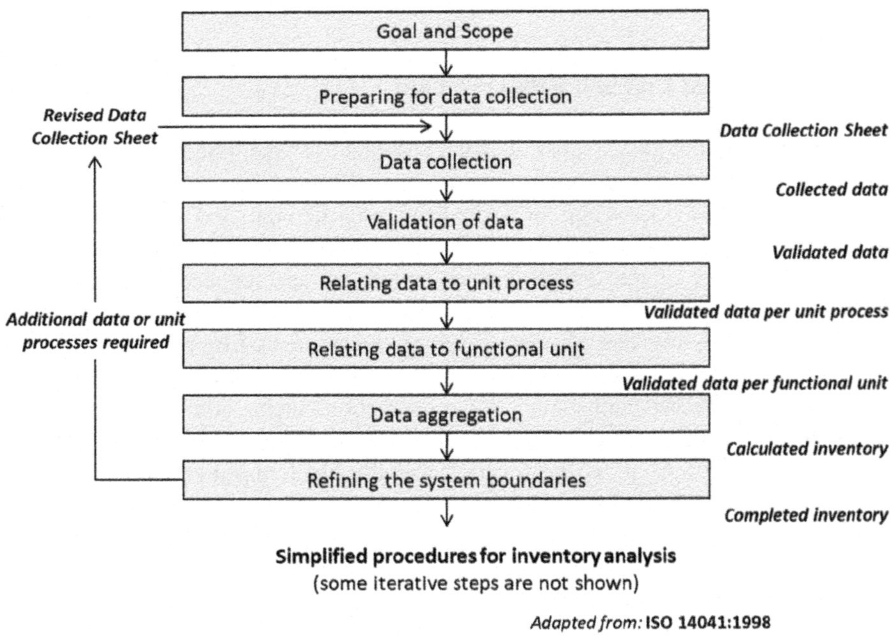

Fig. 6.5 Flow chart for the life cycle inventory analysis

In the collection stage, measurement, calculation and estimation of data related to the unit process are performed using two approaches. In the first approach, data collections performed internal audits at the waste management units. This type of data is known as primary data. Availability of data, finance and time are the significant constraints for this approach. In the second approach, information is collected from the reports and published references available from the previous studies. This type of data is known as secondary data. Data availability and accuracy are the constraints for this approach. Missing data and irregularities is a common problem associated with both the procedures. Data collection based on reports and literature, calculations form the known sources, estimating the value based on experience or accepting and documenting the data gaps are in practise to fulfill the missing data. Quantitative and qualitative characterization of the data is essential to understand the quality of the collected data. The vital factors determining the quality of the data are the geographical coverage (location), temporal coverage (age), and technology coverage. Precision, completeness and representativeness are the additional data requirements to be considered for the data collected from a specific site.

In the validation stage, the data collected is verified by the competent technical people from academic and field to ensure the quality of the data. During the validation, the relation between the data collected, goal and scope, functional unit or reference flow and unit process has to be verified. The calculation procedures implemented should substantiate to confirm the consistency throughout the system boundary.

2.4 *Life Cycle Impact Assessment*

As per ISO 14042:2000a, life cycle impact assessment is defined as "phase of life cycle assessment aimed at understanding and evaluating the magnitude and significance of the potential environmental impacts of a product system". The objective of the impact assessment is to examine the product system using impact categories and category indicators from an environmental perspective and further provide information for the interpretation phase. In this stage, data is converted into pre-defined impact categories for effective communication to the stakeholders. The framework of LCAI includes mandatory and optional elements. The necessary components include the selection of impact categories, assignment of LCI results to impact types (classification) and calculation of impact category indicator results (characterization). The optional elements include normalization, weighting, grouping, and data quality analysis.

The LCA framework (ISO 14044:2006b) does not provide or recommend any impact categories and indicators list. However, the categories, indicators and characterization models intended to be studied should be accepted internationally or authorized by an international board. The persons performing the LCA study determine these selections of components. These are defined during goal and

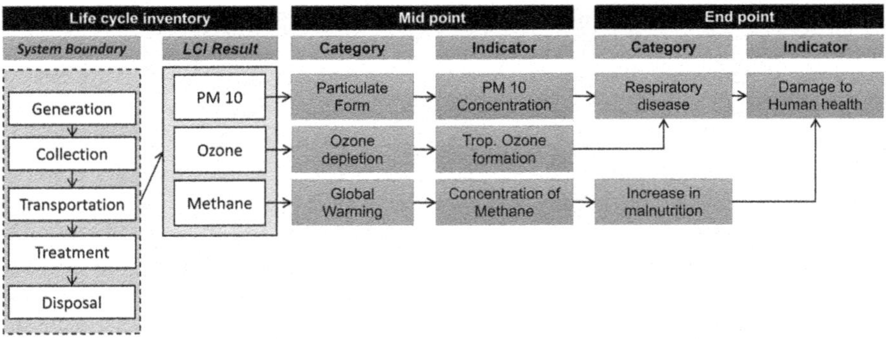

Fig. 6.6 Model of impact assessment structure used in ReCiPe 2016

scope, enabling the user to procure specific data, to fulfill the requirements of the impact assessment.

Classification is a systematic scientific approach in which the input and output data collected during the inventory phase are identified and assigned to the impact categories (environmental impacts). During the process of mapping, the data can be categorized under multiple impact categories. The quantification of the effects associated is acceptable only if the results are independent; else measures are to be taken to avoid the duplication. The classification of the impact categories is based on the spatial and temporal variations. Spatially they are classified as global, continental, national, regional and local levels. The temporal classification plays a pivotal role since some impact categories will have long-term impacts (example global warming). Figure 6.6 shows model of impact assessment structure used in ReCiPe 2016.

Characterization is a systematic quantification of the impacts associated with each category using scientific analyses and models. In this process, the potential effects of the input and output results are determined. They are converted to a common indicator (unit) based on the classification; for example, under global warming indicators all the results are converted to carbon dioxide equivalents. The translated results are summed up to determine the total indicator value.

2.5 *Life Cycle Interpretation*

As per ISO 14043:2000b, life cycle interpretation is defined as "phase of life cycle assessment in which the findings of either the inventory analysis or the impact assessment, or both, are combined consistently with the defined goal and scope to reach conclusions and recommendations". In the interpretation phase, significant issues related to the results obtained from the inventory analysis and impact assessment phase are identified. Further, evaluation is performed in a systematic approach through completeness, sensitivity and consistency checks. Finally, the conclusions,

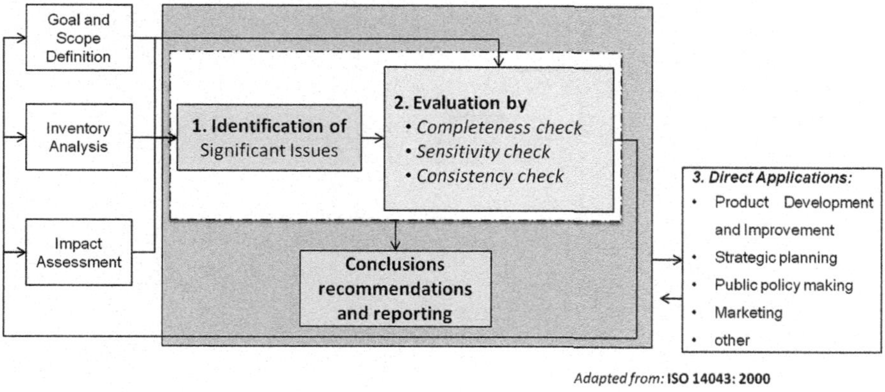

Fig. 6.7 The flow components of the life cycle interpretation stage

recommendations and limitations are documented in a complete, consistent, understandable and transparent manner. Figure 6.7 represents the flow components of the life cycle interpretation stage.

3 Case Studies

3.1 *Kerbside Organics Disposal and Treatment, Auckland Council*

This case study illustrates the life cycle assessment performed by Auckland Council to assess the environmental impacts from landfill of kerbside organics (KSO) compared to composting, anaerobic digestion (AD), and the combination of AD and composting. The primary emphasis of this case study is to illustrate that the LCA can be performed using the mathematical calculations in the absence of the software. This type of analysis can form a basis for decision making during the preliminary assessment stages (Dubey and Singhal 2014).

The organic component of the kerbside collection refuses stream constitutes 40% food waste (FW) and 10% garden waste (GW). To achieve this, following four scenarios are compared: Landfilling of KSO material assuming 90% gas recovery efficiency; Aerobic composting of KSO and application of a product to land; Anaerobic digestion of KSO with energy recovery and application of digestate to land; and Anaerobic digestion of KSO followed by aerobic composting of digestate before application to soil. The life cycle inventory database was collected from secondary sources (i.e., literature and previous studies). The research is focused on performing mass balances for carbon, nutrients (nitrogen), and global warming potential (GWP). Figure 6.8 illustrates the scenarios considered for performing the life cycle assessment.

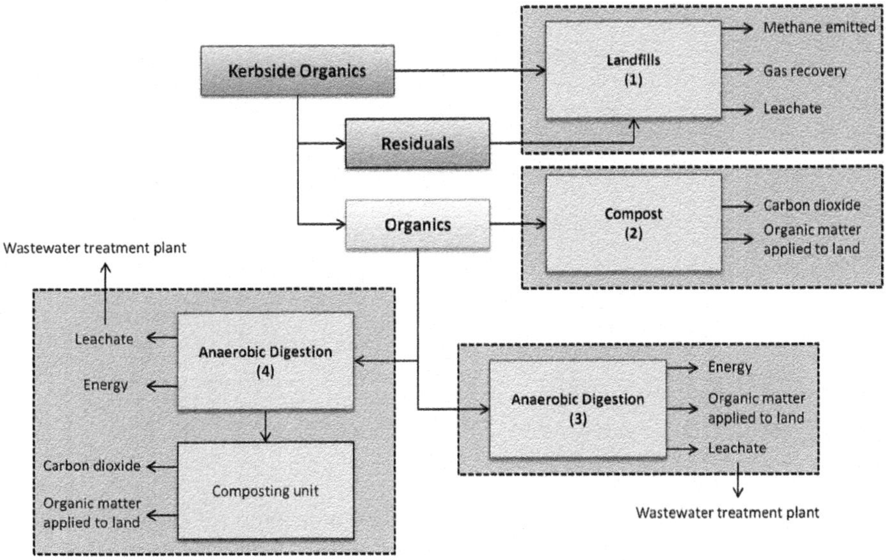

Fig. 6.8 Scenarios for disposal and treatment of kerbside organics

The major assumptions that are made during this study are:

1. In the council 248,590 tons of KSO are produced each year and 40% of KSO is FW and 10% is GW.
2. In the landfill scenario (1), 90% of gas produced is captured; 76 kg of CH_4 and 180 kg of CO_2 are produced per each ton of FW, and 99 kg of CH_4 and 232 kg of CO_2 are produced per each ton of GW; 1% of total carbon is lost in leachate.
3. In the Composting Scenario (2), 1 Kg of CH_4 and 115 Kg of CO_2 Are Produced per each Ton of FW and 3 Kg of CH_4 and 194 Kg of CO_2 Are Produced per each Ton of GW; 350 Kg/Ton-FW and 200 Kg/Ton-GW of Carbon Are Retained in Compost

   - In the anaerobic digestion scenario (3), 70 kg of CH_4 and 50 kg of CO_2 are produced per each ton of FW and 50 kg of CH_4 and 30 kg of CO_2 are produced per each ton of GW; 0.6 ton of digestate is produced per each ton of organic waste.
   - In the anaerobic digestion followed by composting of digestate scenario (4), 52 g of CH_4 is assumed to be produced per ton of digestate.

A sample mass balance calculations for the carbon mass balance in the landfill scenario is presented in Table 6.1.

The carbon mass balance study was performed for the four scenarios using the mathematical calculations in an excel spreadsheet. Table 6.2 presents the mass carbon balance for all the scenarios under consideration. For the landfill scenario, 46,219 tons/year of carbon (78% of carbon in food and garden waste deposited in landfills) is lost via retention in the landfill. In comparison, composting, anaerobic

Table 6.1 Mass balance calculations for the carbon mass balance in the landfill scenario

		Wet waste				Carbon			
Qualifier	Parameter	Basis	Unit	Amount	Unit	Basis	Unit	Amount	Unit
Input	KSO			248,590	Tons/year				
Input	*Food waste (FW)*	40	% of KSO	99,436	Tons/year	48	%	47,729	Tons-C/year
Output	CH_4 produced	76,622	g/ton	7619	Tons/year	0.75	g-C/g-CH_4	5714	Tons-C/year
Output	CH_4 released	10	% biogas produced	762	Tons/year	0.75	g-C/g-CH_4	571	Tons-C/year
Output	CH_4 captured	90	% biogas produced	6857	Tons/year	0.75	g-C/g-CH_4	5143	Tons-C/year
Output	CO_2 released	180,183	g/ton	17,917	Tons/year	0.27	g-C/g-CO_2	4886	Tons-C/year
Input	*Garden waste (GW)*	10	% of KSO	24,859	Tons/year	47.8	%	11,883	Tons-C/year
Output	CH_4 produced	98,514	g/ton	2449	Tons/year	0.75	g-C/g-CH_4	1837	Tons-C/year
Output	CH_4 released	10	% biogas produced	245	Tons/year	0.75	g-C/g-CH_4	184	Tons-C/year
Output	CH_4 captured	90	% biogas produced	2204	Tons/year	0.75	g-C/g-CH_4	1653	Tons-C/year
Output	CO_2 released	231,664	g/ton	5759	Tons/year	0.27	g-C/g-CO_2	1571	Tons-C/year
MB							Total C released (g)	14,008	Tons-C/year
Output	Leachate C released	1%					Total C loss	140	Tons-C/year
C lost via retention in landfill + emissions							78%	46,219	Tons-C/year

C lost via retention in landfill + emissions = [Input Food waste + CH_4 released (FW) + Input Garden Waste + CH_4 released (GW)] – [Total C released + Total C loss]

% C lost via retention in landfill + emissions = C lost via retention in landfill + emissions/[Input (FW + GW)]

Table 6.2 Carbon mass balance results for the scenarios under consideration

Scenario	Mass (Tons-C/year)	Fraction of input (%)
Landfill	46,219	78%
Compost	4108	8%
Anaerobic digestion (AD)	2594	5%
AD with composting of digestate	2597	5%

digestion, and anaerobic digestion followed by compositing result in 4108 tons/year (8%), 2594 tons/year (5%), and 2597 tons/year (5%) of carbon lost from the system.

3.2 Comparison of Organic Processing Odour Control Technologies

A comparative LCA of odour control technologies used in organic processing units was performed by Bindra (2015), and his team was illustrated in this case study. Three odour control technologies packed-bed wet scrubber (PBWS), organic bio-filter system (OBFS) with wood chips media and inorganic bio-filter system (IBFS) with synthetic media were compared. The study is aimed to assess the potential environmental burdens of the odour control technologies in organic processing systems. The assessment was performed over a period of 15 years based on the life expectancy of the inorganic media used in these technologies. In the system boundary, raw materials for construction, transportation, energy utilization, water consumption, and chemical usage, recycling, and disposal phases were considered. Figure 6.9 illustrates the system boundary and the process considerations for the three technologies.

The life cycle inventory database was developed by performing field surveys at the organic processing plants to understand the technologies and collect the preliminary data. The data for OBFS with wood chip media was obtained from a composting plant located in Guelph; the processing capacity of the plant is 30,000 tons of organic waste a year. The data for the IBFS with synthetic media was collected from a centralized composting facility located in Hamilton; the processing capacity of the plant is 70,000 tons of organic waste a year. The inventory database for the PBWS system was developed from the secondary sources (i.e., literature and previous studies). For the analysis and comparison of the technologies, 1000 t of organic waste per year was selected as functional unit. Life cycle inventory database was developed by conducting the field visits at both the compost units. For all the three technologies, the components contributing in significant quantity were considered (chemicals, equipment, etc.), while small parts (screws, bolts, wires, etc.) with negligible effect were excluded. The data gaps were quantified and scaled to the functional unit based on the reasonable assumptions from the background of the system, field data, and common knowledge. SimaPro® V.8.0 software and Ecoinvent v3.0 databases were used to perform impact analysis. Twelve environmental impacts

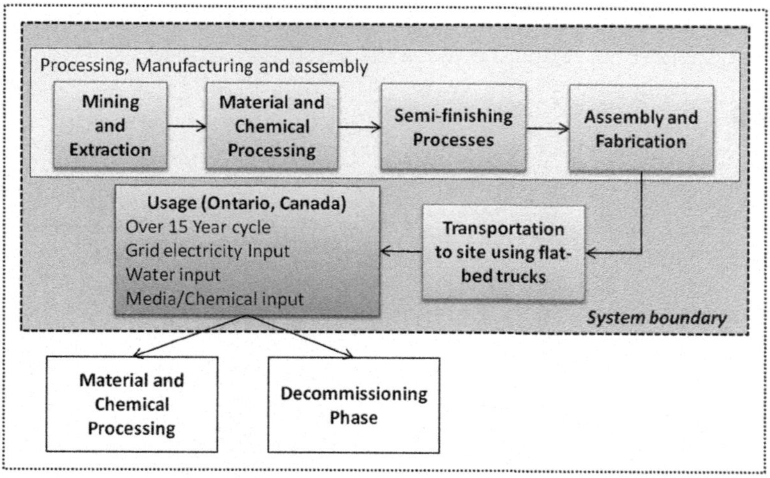

Fig. 6.9 System boundaries for the odour control technology

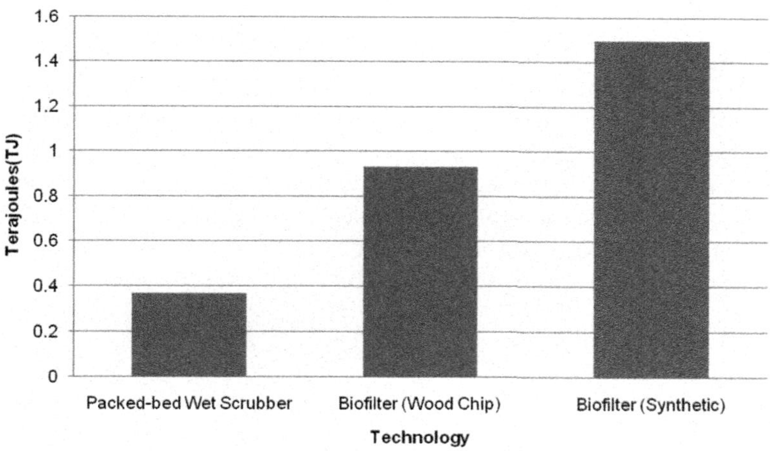

Fig. 6.10 Cumulative energy demand for each technology. (Adopted from Bindra et al. 2015)

categories were considered as a part of this study. The categories include non-toxic impact categories include cumulative energy demand (CED), climate change, freshwater, and marine eutrophication; Toxic impact categories include human, terrestrial, freshwater, and marine eco-toxicity, photochemical oxidant formation, terrestrial acidification; Resources include metal depletion, fossil depletion. The source mix of the Ontario city electricity generation for the year 2014 was considered to determine the CED. The sources of the electricity generation include nuclear (62%), hydro (24%), gas/oil (10%) and wind (4%). Figure 6.10 represents the CED for each odour control technology.

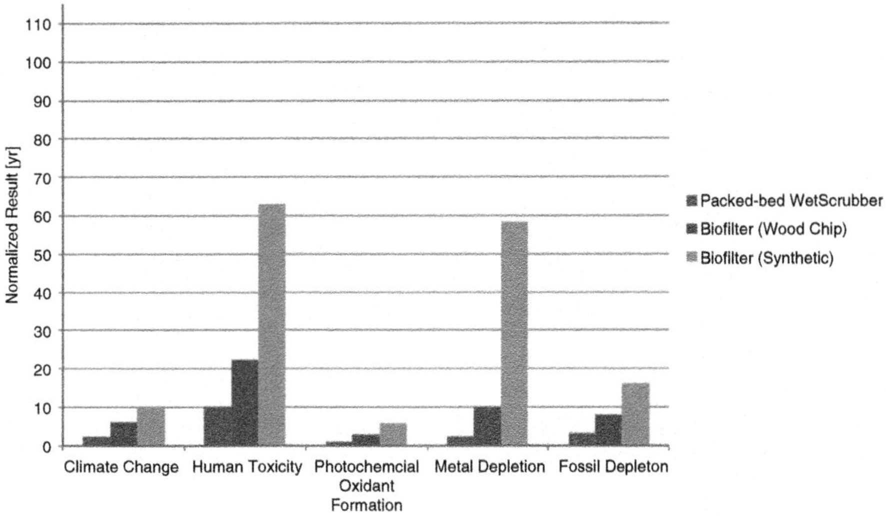

Fig. 6.11 Comparative normalized results of selected impact categories for odour control technologies. (Adopted from Bindra et al. 2015)

For the PBWS the total demand of CED is 0.37 TJ; for the OBFS with wood chip media the total demand of CED is 0.93 TJ, and for the IBFS with synthetic media CED is 1.5 TJ. The normalized impact assessment results for the non-toxic impact categories (climate change), toxic impact categories (human toxicity, photochemical oxidant formation), and resources (metal and fossil depletion) are presented in Fig. 6.11. Based on the LCA study, impact associated with the PBWS is the lowest for all the impact categories followed by the OBFS with wood chip media. The environmental impacts are highest for the IBFS with synthetic media.

References

Bindra, N., Dubey, B. and Dutta, A. (2015). Technological and Life Cycle Assessment of Organics Processing Odour Control Technologies. *Science of the Total Environment*, **527–528:** 401-412.

Dubey, B. and Singhal, N. (2014). Organics Waste Diversion from Landfills – Case Study from Auckland, New Zealand. *Clean India Journal*, July-Sep 2014, 46–48.

ISO (International Standard Organization) (1997). ISO 14040. Environmental Management—Life Cycle Assessment: Principles and Framework. CEN European Committee for Standardization, Brussels.

ISO (International Standard Organization) (1998). ISO 14041. Environmental Management—Life Cycle Assessment: Goal and Scope Definition and Inventory Analysis. International Standard Organization for Standardization, Geneva.

ISO (International Standard Organization) (2000a). ISO 14042. Environmental Management—Life Cycle Assessment: Life Cycle Impact Assessment. International Standard Organization for Standardization, Geneva.

ISO (International Standard Organization) (2000b). ISO 14043. Environmental Management—Life Cycle Assessment: Interpretation. International Standard Organization for Standardization, Geneva.

ISO (International Standard Organization) (2006a). ISO 14040:2006, ISO/TC 207/SC5. Environmental Management—Life Cycle Assessment: Principles and Framework. International Standard Organization for Standardization, Geneva.

ISO (International Standard Organization) (2006b). ISO 14044:2006, ISO/TC 207/SC5. Environmental Management—Life Cycle Assessment: Requirements and Guidelines. International Standard Organization for Standardization, Geneva.

Klöpffer, W. and Grahl, B. (2014). Life Cycle Assessment (LCA). A Guide to Best Practice. Wiley-VCH, Weinheim.

Sankar Cheela, V.R. and Dubey, B. (2019). Review of Application of Systems Engineering Approaches in Development of Integrated Solid Waste Management for a Smart City. *In:* Rathinasamy, M., Chandramouli, S., Phanindra, K. and Mahesh, U. (eds). Water Resources and Environmental Engineering II. Springer, Singapore.

SETAC—Society of Environmental Toxicology and Chemistry (1993). Guidelines for life-cycle assessment: A 'Code of Practice'. From the SETAC Workshop held at Sesimbra, Portugal, 31 March–3 April 1993. Edn. 1, Brussels and Pensacola (Florida), August 1993.

Chapter 7
Reuse and Recycle: A Green Chemistry Approach

Ashish Kumar Nayak and Anjali Pal

1 Introduction

Green chemistry is a sustainable approach in which the substance is to be synthesized in an accurate, non-polluting and protected manner. Also it requires least quantity of energy and resources with minimal generation of waste materials. It is necessary to diminish the impairment of nature by anthropogenic materials, and the processes employed to generate them.

Green chemistry signifies that the research about the effluence responsiveness comes out from scientific innovations. It involves twelve set of principles which minimize or prevent the use or production of unsafe substances. The principles of green chemistry can be accomplished by implementing reproducible, harmless and environmental friendly efforts during the production of chemical products, and in researches. In regard, scientists can significantly reduce the plausible risks to the environment and human health with the help of valuable ideologies of green chemistry. The primary goals of green chemistry are mentioned here.

1. It helps to build up the technical dimension of a sustainable civilization.
2. It aims to identify, understand and replace unsustainable products and processes with those sustainable alternatives.
3. It aims to develop a field of chemistry that can replace the polluting activities either in the shape of products or processes.

A. K. Nayak · A. Pal (✉)
Civil Engineering Department, Indian Institute of Technology Kharagpur, Kharagpur, West Bengal, India
e-mail: anjalipal@civil.iitkgp.ac.in

D. Sengupta et al. (eds.), *Treatment and Disposal of Solid and Hazardous Wastes*,
https://doi.org/10.1007/978-3-030-29643-8_7

2 Green Chemistry: A Sustainable Methodology

Green chemistry is a philosophy of chemical research and engineering that encourages the design of chemical processes and products, which acquires sustainable, harmless and non-polluting characteristics. It also consumes minimum quantity of energy and materials while generating little or zero-waste materials. There are several important aspects in which green chemistry is considered to be sustainable:

- *Materials:* By proficiently using renewable materials, maximum recycling and minimum use of pristine raw materials, the green chemistry is sustainable with respect to materials.
- *Economic:* At high level of superiority, green chemistry usually expenses less in firmly economic conditions than chemistry as it is normally accomplished.
- *Waste:* By minimizing so far as possible or even totally exterminating their production, we can conclude that green chemistry is sustainable with respect to wastes.

3 Twelve Fundamental Principles of Green Chemistry

Anastas and Warner (1998) had developed the twelve principles of green chemistry as it gives instructions to expert chemists to implement new substance, modern production and modern scientific processes. These principles are given below:

1. *Prevent waste:* The process is designed in such a way that the waste is prevented at the beginning stage rather than cleaning or treating up the waste after it has been produced.
2. *Maximize atom economy:* During the formation of chemical products, synthetic methods should be proposed in a way to maximize the incorporation of all starting materials used in the final product.

The concept of atom economy also explores the quantity of the unwanted product evolved in a particular reaction. It is defined as follows:

$$\text{Atom economy} = \frac{\text{Molecular weight of desired product}}{\text{Sum of molecular weight of all the reactants used}}$$

The above definition is taken as a measure of how the raw material is used efficiently. Raw material resulting maximum atom efficiency is to be chosen.

For example, maleic anhydride ($C_4H_2O_3$) can be generated by employing either *n*-butane (C_4H_{10}) or benzene (C_6H_6) as the reactant species and the corresponding chemical reactions are expressed as:

$$C_4H_{10} + \frac{7}{2}O_2 \rightarrow C_4H_2O_3 + 4H_2O \quad (\text{form } n\text{-butane})$$

$$\%\text{Atom economy} = 100 \times \frac{\text{Mass of }(4C + 2H + 3O)\text{atoms}}{\text{Mass of }(4C + 10H + 7O)\text{atoms}}$$

$$= 100 \times \frac{98\,\text{amu}}{170\,\text{amu}} = 57.6\%$$

$$2C_6H_6 + 9O_2 \rightarrow 2C_4H_2O_3 + 4H_2O + 4CO_2(\text{form benzene})$$

$$\%\text{Atom economy} = 100 \times \frac{\text{Mass of }(8C + 4H + 6O)\text{atoms}}{\text{Mass of }(12C + 12H + 18O)\text{atoms}}$$

$$= 100 \times \frac{196\,\text{amu}}{444\,\text{amu}} = 44.1\%$$

Atom economy for the *n*-butane step is 57.6% and that for the benzene step is 44.1%. Therefore, *n*-butane is favoured over benzene as a raw material for the production of desirable product—maleic anhydride.

3. *Design less hazardous chemical synthesis:* This incorporates the development of manufacturing strategies that will reduce the use and formation of toxic and hazardous substances. Also, it generates the eco-accommodating substance that has no dreadful effect on human health and the environment.
4. *Design safer chemical products:* Chemical products to be employed in different activities should have the efficacy to do their job with minimum harm to the environment as well as to the human health.

For example, adipic acid ($HO_2C(CH_2)_4CO_2H$) is broadly employed in polymer industries for the manufacturing of nylon, polyurethane, etc. In this regard, benzene is being used for the synthesis of adipic acid which is carcinogenic in nature. But it can also be synthesized by simply adopting the green technology invented by Drath and Frost (1994), in which adipic acid is enzymetically prepared from the glucose molecules.

5. *Use safer solvents and auxiliaries:* It aims to utilize green solvents (e.g. water, supercritical CO_2) rather than volatile organic solvents (e.g. liquor, benzene known for cancer-causing, CCl_4, $CHCl_3$, perchloroethylene, CH_2Cl_2) for the synthesis of chemical products and other activities. If needed, it is better to use the solvents that are environment friendly.
6. *Design for energy efficiency:* During the synthesis of a chemical product, the energy required to carry out a reaction should be minimized in order to make the process more cost-effective. If possible, the synthetic processes should be conducted at ambient temperature and pressure conditions.
7. *Use renewable feedstock:* It encourages the use of raw material or a feedstock which should be renewable rather than depleting, wherever economically and technically possible. Otherwise, the use of sustainable or renewable resources

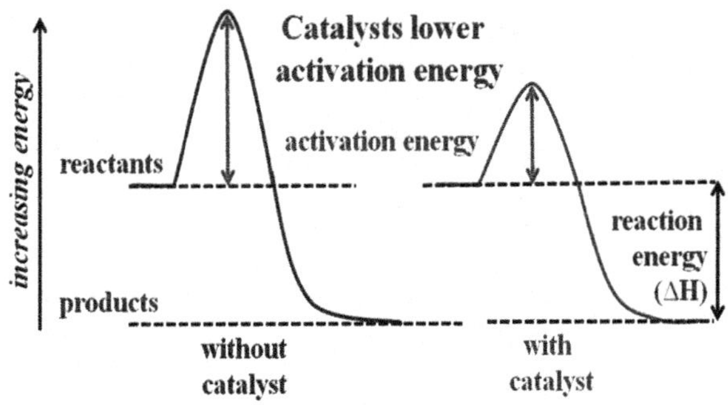

Fig. 7.1 Lowering the activation energy of a reaction by employing a catalyst

does not put much nuisance on the environment. The end-products and wastes are also biodegradable in nature.

For example, the biochemical synthesis of adipic acid using glucose as a feedstock is an eco-friendly substitute for the conventional chemical synthesis of adipic acid using benzene as the starting material.

8. *Reduce the superfluous derivatization:* The unnecessary steps like protection/deprotection of groups, blocking of group, modification of physical and chemical processes, etc. should be minimised, if possible (wherever technically and economically practicable). These strides can lead to the requirement of additional reagents and minimize the production of waste materials.
9. *Use catalysis in place of stoichiometric reagents:* Catalyst-based synthesis of chemical products cause less pollution generation compared to synthesis routes that are derived from stoichiometric reagents.

Therefore, a catalyst plays a crucial role in the designing of eco-friendly chemical processes because of the following advantages:

(i) The catalysts are substances that increase the reaction rate of a chemical reaction without being consumed in the process. It helps to lower the activation energy for the process (Fig. 7.1).
(ii) Maximum utilization of the starting material (or feedstock) and minimum generation of waste products.
(iii) The ideal catalysts are fully recovered after the reaction without any change in physical or chemical properties which results to 100% atom economy.
(iv) This will improve the selectivity towards reaction products and plausibly keep away from unwanted side responses prompting an immaculate innovation with maximum yield capacity.

Therefore, a significant improvement in waste reduction can be accomplished through proper selection and designing of reaction catalysts.

10. *Design chemicals and products which degrade after use:* Chemical products should be designed in such a way that they do not persist in the environment

after they are utilized, or till they break down easily into nnocuous degradation products.

For example, biopolymers such as polyhydroxyalkanoates (PHA) and polylactic acid (PLA) are employed as substitutes for chemical plastics which exhibits excellent biodegradable features as compared to plastics.

11. *Real-time analysis for pollution prevention:* Analytical methodologies should be built up to permit the real-time in-process supervising and interruption prior to the formation of hazardous substances.

It explores the process at which the chemists can monitor the progress of chemical reactions completely. By simply identifying when the product will be ready, it is possible to save a lot of energy and time, and reduce the waste. This efficient supervising is also very essential to prevent accidents.

12. *Inherently safer chemistry for accident prevention:* Such substances and their derivatives are to be chosen which can be employed in the chemical processes so that they can minimize the risk of chemical accidents including fire and explosions.

For example, if the chemical processes take place with the gaseous substances, the possibility of chemical accidents might be higher as compared to the processes working with the liquid or solid substances. On the other hand, the risk would be minimum if the chemical processes work with solid substances.

We all know that the Bhopal gas tragedy happened on 3rd December 1984, and is considered to be the worst industrial disaster in history. It was reported that 40 tons of methyl isocyanate (MIC) gas was accidentally leaked when a holding tank was overheated at the Union Carbide India Limited pesticide plant, located in the heart of the city of Bhopal (Madhya Pradesh), India. Nearly 15,000 people died and more hundreds of thousands were injured. Thus, chemists seek to avoid things that explode, light on fire, air-sensitive, etc.

4 Advantages of Green Chemistry

Green chemistry is significant as it leads to various important environmental benefits and much more prosperous economy. These advantages are given below:

Business benefits of green chemistry include:

1. It enhances the resource efficiency and minimum raw material, and utility bills.
2. It diminishes the waste treatment and disposal costs.
3. It has the ability to capitalize on the environment technology marketplace by enforcing new 'greener' practices.
4. It improves the health and safety of staff and customers.

Environmental benefits include:

1. Less quantity of raw materials and natural resources are to be utilized.
2. It releases the lower levels of chemicals into the environment.
3. It invents cleaner production technologies.
4. It reduces the emissions as well as product impacts.
5. Minimal amount of hazardous waste is to be treated and disposed off.

5 Applications of Green Chemistry and Technology for Waste Minimization and Resource Recovery

5.1 Surfactant Adsorption on Solid Surfaces and Its Further Application in Dye Removal

The surfactants are well known surface-active agents that can decrease the surface tension of water effectively. These surface-active species contain both strong hydrophilic (water-soluble) and hydrophobic (water-insoluble) components. The hydrophobic group is usually a hydrocarbon chain consisting of 10–20 carbon atoms. The hydrophilic group can be ionized depending on the surfactant category. At lower levels of concentration, the surfactants perform similarly as the normal electrolytes, whereas at higher concentrations they behave differently. Due to the configuration, the surfactant molecules align themselves in water to form large aggregates called micelles. In a micelle, the hydrophobic tails flock in the interior portion of the aggregate in order to minimize their contact with water, whereas the hydrophilic heads remain on the exterior surface in order to maximize their contact with water (Fig. 7.2). In general, the micellization process in water is decorated as an agreement between the affinity of alkyl chains to repudiate energetically unfavourable interactions with water, and the surfactant polar heads are enforced to maintain contact with the aqueous environment. The concentration of a surfactant in bulk phase above which aggregates of tenside molecules are developed is known as critical micelle concentration (CMC). These micelles have the stupendous ability to absorb the organic solutes from solution, and this process is called solubilization (Fig. 7.3). This solubilizing phenomenon of surfactants makes them very useful for domestic and industrial applications.

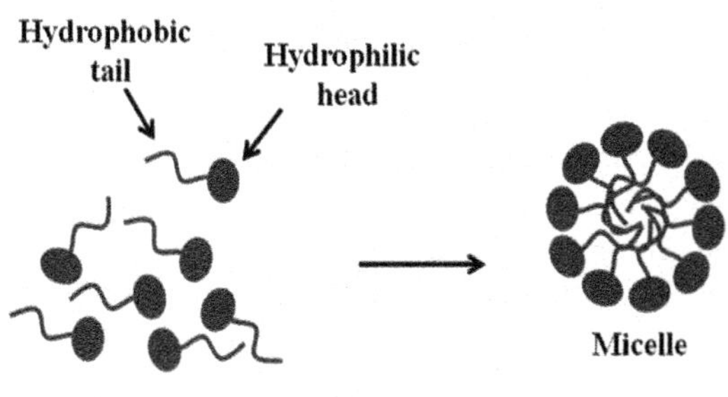

Fig. 7.2 Surfactant molecules and micelle formation above critical micelle concentration (CMC)

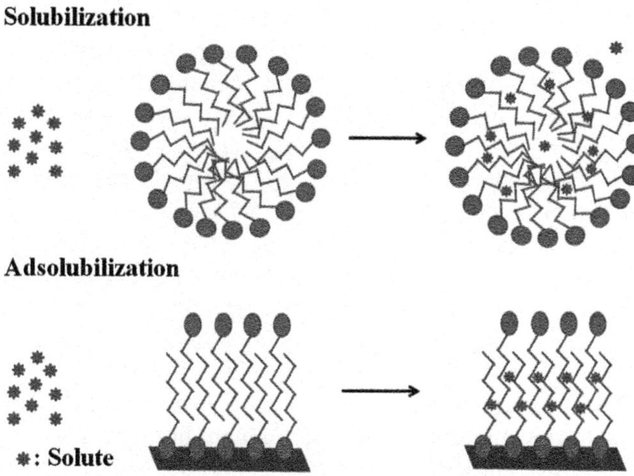

Fig. 7.3 Schematic diagram of solubilization and adsolubilization phenomena

The surfactants are classified into four groups depending upon the nature of the hydrophilic heads: cationic, anionic, non-ionic and zwitterionic (amphoteric). Cationic and anionic surfactants together are labelled as ionic surfactants. Ionic surfactants represent approximately two-thirds of all the surfactants, whereas anionic surfactant possesses more than 90% of the ionic constituents alone (Ayranci and Duman 2007). The predominant examples of anionic surfactant are sodium stearate, sodium dodecyl sulfate, sodium dodecyl benzene sulfonate, dialkyl sulfosuccinate, etc.

In recent years, the production of detergent is based largely on linear alkyl sulfate such as sodium dodecyl sulfate. These surfactant-based detergents are being employed extensively in households for washing and cleaning purposes. It has been reported that the worldwide production of surfactant quantities increased exponentially (which is >10 million tons per year) and nearly 60% of its production is measured as anionic surfactants only (Schouten et al. 2007). The anionic surfactants have also been employed in industrial activities such as textiles, fibres, cosmetics, paints, food, polymers, pharmaceuticals, mining, oil refinery and pulp-paper industries. These applications of anionic surfactants may enhance its concentrations in the effluents which lead to huge foam formation, and they also penetrate into the underground water resources.

Surfactants are harmful to human beings, fish and vegetation, and are responsible for causing foams in rivers, effluent treatment plants and also to diminish the quality of water. Surfactants cause long-term as well as short-term changes in the ecosystem. Based on the above-mentioned reasons, most of the public health and environmental regulatory agencies had maintained stringent limits for anionic surfactant as 0.5 mg/L for drinking water and permissible up to 1.0 mg/L for other activities (Adak et al. 2005). Thus, appropriate treatment methodologies are required that can eradicate the surfactant contamination from domestic and industrial effluents.

Surfactant removal operations from the aqueous environment include methods such as chemical oxidation, adsorption, chemical precipitation, reverse osmosis, biological methods etc. In most of the cases, the surfactant concentrations were reported in the same range as that of domestic wastewater effluents whereas the industrial operations like textile, oil refinery, tannery, etc. are discharging the surfactant wastewater at the scale of >1000 mg/L. Also, the surfactant concentration of pond water in rural areas has been accounted at the extreme levels because various washing and bathing activities happen in it. Among the aforesaid treatment methods, many practices are not environment-friendly or cost-effective and also not feasible for functioning in household scale. In this regard, the adsorption techniques are economical and can be implemented in local households, even in low and middle-income groups.

In the first step, the adsorption study aims to remove the anionic surfactant from the high-strength synthetic solutions using alumina. In the second stage, the reusability potential of the surfactant-modified alumina (SMA) has been examined for the removal of cationic dyes as well as heavy metals from the aqueous environment so that the surfactant-loaded alumina can be treated in a sustainable way. The detailed procedures are mentioned below.

Step 1. Adsorption of Anionic Surfactant on Alumina

The adsorption behaviour of anionic surfactant on positively charged alumina surface is completely different from ideal adsorption process. Here, the anionic surfactants are assembled onto the positively charged alumina surface resulting in terms of monolayer or bilayer shape which depends upon the initial surfactant concentration. At low surfactant concentration, the surfactant molecules are adsorbed independently whereas, at high adsorbate concentration, the monolayer and bilayer structures are produced. These structures behave as micelles and also have the ability to solubilize the organic molecules into their three-dimensional structures. The phenomenon is called adsolubilization (Fig. 7.3). The monolayer structure is known as hemimicelle while the bilayer structure is called admicelle. After the removal of anionic surfactant under high concentration conditions, the obtained SMA contains similar arrangements like that of admicelles, and further can be utilized for pollution remediation. Sodium dodecyl sulfate (SDS) has been considered as a representative member of anionic surfactant for this study. The procedure reported earlier by Adak et al. (2005) is chosen to prepare SMA (Fig. 7.4).

Surfactant Adsorption Isotherm

The adsorption study of an ionic surfactant onto the oppositely charged solid surface followed a four-regime isotherm (Fig. 7.5), which are described below.

- **Region I** adsorption governed by Henry's law. Here the surfactants are adsorbed as monomers, and they do not interact with each other. The adsorption phenomenon occurs primarily from electrostatic forces between surfactant ions and the oppositely charged solid substrate.
- **Region II** illustrated a sudden increase in the slope of adsorption isotherm. In this region, the adsorption takes place due to the electrostatic attraction between the surfactant ions and the oppositely charged solid surface, and hemimicelles

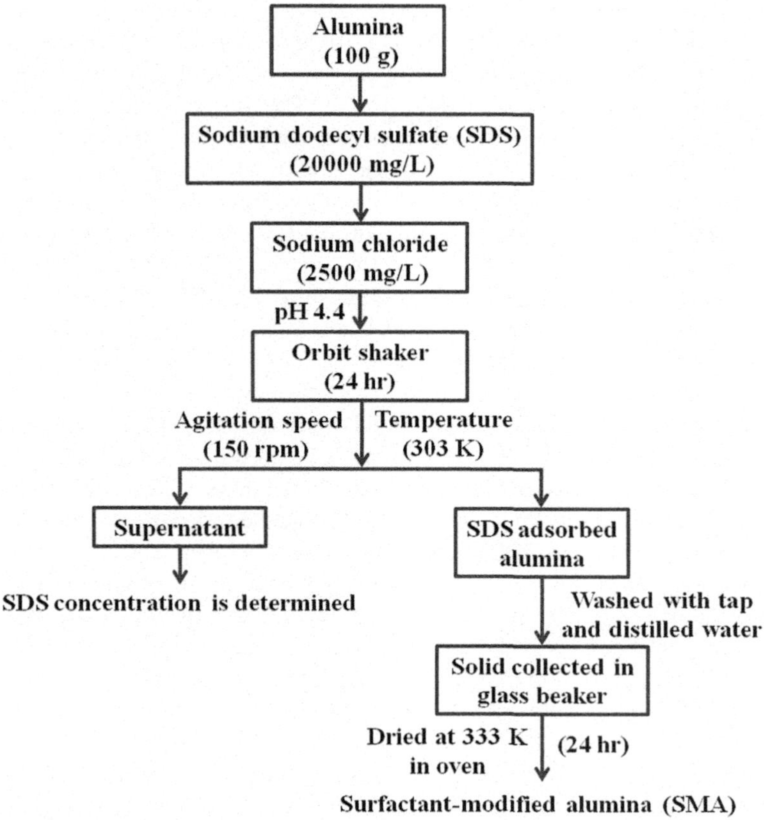

Fig. 7.4 Scheme showing the proposed steps for obtaining SMA material

Fig. 7.5 Schematic diagram of a typical surfactant adsorption isotherm

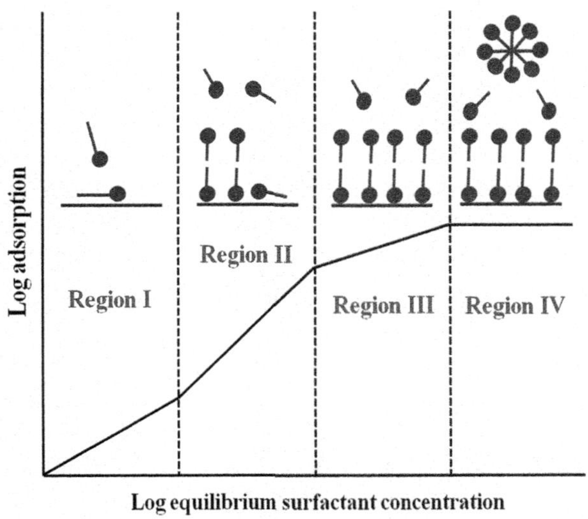

association of hydrocarbon chains. Also, micelle-like aggregates are generated on the solid substrate in this region. The concentration at which the transition from the region I to region II happens has been documented as critical admicelle concentration (CAC) or hemimicelle concentration (HMC). The monolayered structure is called hemimicelle whereas the bilayered structure is referred to as admicelle.

- **Region III** depicted a slower rate of increase in adsorption compared to region II. Here the admicellar structures are not fully formed and allow further adsorption. The adsorption in region II is due to the association between the hydrocarbon chains.
- **Region IV** is the plateau adsorption region above the CMC with the formation of fully formed aggregates and saturation levels for surface coverage. The transition from region III to region IV occurs near the CMC of the surfactant.

Therefore an appropriately chosen solid substrate and experimental conditions should be employed for the treatment of surfactant bearing wastewater.

The above-mentioned conditions are incorporated in the SDS adsorption isotherm using alumina as the substrate, and the resulted isotherm has been similarly distinguished into four regions. The region I revealed at an extremely low equilibrium concentration (<3.37 mg/L of SDS). The transition from the region I to region II explored admicelle or hemimicelle concentrations. The transition zone occurred in between region II and region III at an equilibrium concentration of 30.2 mg/L. The transition from region III to region IV carried out at an equilibrium concentration of 8839 mg/L. From this isotherm study, the maximum adsorption capacity is found to be 111.6 mg/g with an initial SDS concentration of 20,000 mg/L.

In this observation, it has been shown how alumina, an abundantly occurring cheap and non-toxic material, can remove the anionic surfactant such as sodium dodecyl sulfate (SDS), when present in water at high concentration. The adsorption isotherm study revealed four distinct regions. The equilibrium time is found to be 2 h. At neutral pH, the removal efficiency is minimum (i.e. ~65%). However, the efficiency can be increased significantly (i.e. >98%) with decreasing the pH value up to 2.0. On the other hand, it can be seen that the removal efficiency is increased with the addition of NaCl solution. After the adsorption of SDS on alumina, the spent alumina can be utilized for the removal of cationic dyes such as malachite green (MG), crystal violet (CV), etc. from the aquatic environment. This process is recognized as 'adsolubilization', as shown in Fig. 7.3. The main interaction between the cationic dye and the negatively charged surface of SMA is ionic. However, hydrophobic interaction between these two also can occur simultaneously.

Step 2. Adsolubilization of MG Dye on SMA

Various classes of dyes seek their application in different industrial sectors like textiles, pulp and paper, cosmetics, paint and varnishes, ink, plastics, tanneries, etc. A significant quantity of dye (~10%) has vanished during the industrial activities which result in the production of large amount of dye bearing wastewater. These coloured dye effluents deteriorate not only the aesthetic water belongings but also enhance the organic loading in it. The elevation of organic loading inhibits the penetration of sunlight into the receiving water bodies and consequently, the photosynthetic activity of the autotrophic organisms is reduced. Also, it creates an ecological imbalance in the surrounding ecosystem. The complex aromatic structures existing in synthetic dyes made them difficult for the anaerobic digestion and are resistant to heat, light and oxidising agents. Most of the synthetic dyes are extremely toxic and even carcinogenic, and this causes a serious hazard to aquatic flora and fauna.

Among various dyes, malachite green (MG) is a well-known cationic dye because of its extensive applications in different industrial and medical activities. On the other hand, MG is known to be toxic towards the mammalian cells, since it has cytotoxic properties, because of which the removal of MG dye is very essential in the treatment of contaminated wastewater. In addition, synthetic dyes and surfactants are the two contaminants occurring together in many industrial effluents, even at high concentration. It is, therefore, very important to find methods for the removal of surfactants and dyes from the aqueous environment when present at high concentration.

A notable execution is the work of Das et al. (2009a). In their study of the adsorption of MG dye to SMA, they suggested that the pristine alumina was not much efficient for dye remediation. This led to the proposal of a pH_{ZPC} concept for adsorption. From the experimental data, the point of zero charge (pH_{ZPC}) for alumina was found to be 9.15. This reveals that below this pH the alumina surface was positively charged and hence MG being a cationic dye had no attraction for the alumina substrate. On the other hand, when the surface of alumina was modified with SDS molecules to configure admicelle structures on it with anionic heads remain at the top, after that it became more responsive towards the adsorption of cationic dyes to its surface.

In case of MG adsorption study, the pristine alumina (at a dose of 10 g/L) was shaken for 4 h at 150 rpm with solution pH of 5.3. During the experiment, only ~13% of the MG could be removed (Das et al. 2009a, b). However, SDS modification increased the removal efficiency to a large extent as compared to the unmodified alumina for MG adsorption in batch mode. The process was very fast and within 10 min ~97% removal of MG was observed with SMA at a dose of 10 g/L. The schematic diagram for SMA preparation and removal of MG dye by SMA are shown in Fig. 7.6.

The isotherm models and kinetic behaviour of MG dye adsorption onto SMA were investigated. The equilibrium data followed the Langmuir isotherm model with maximum adsorption capacity of 185 mg/g. The adsorption kinetics of MG removal using SMA explored in 1 h contact time was sufficient to reach the equilibrium. The effect of solution pH on adsorption of MG by SMA was investigated in the pH range between 2.5 and 5.75 at a constant dose of 2 g/L and 20 mg/L is the initial MG concentration. It was observed that MG adsorption efficiency enhances with the increase of pH up to 4.75 and then decreases gradually. The maximum removal efficiency of MG was observed at pH 4.75.

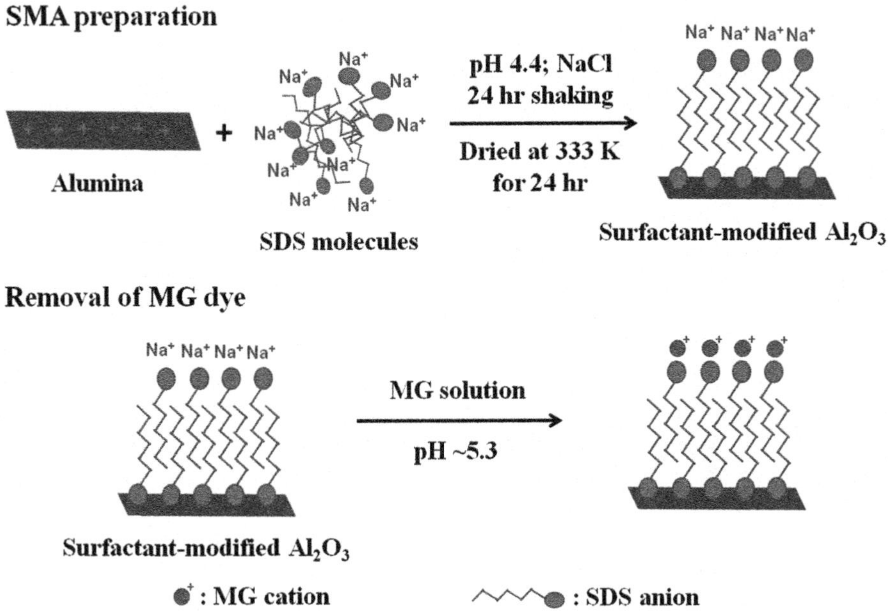

Fig. 7.6 Scheme of adsorptive behaviour of SMA on MG dye

Regeneration of SMA

Recovery and regeneration of exhausted adsorbent will reduce environmental load. For this purpose: 10 g/L dose of SMA was mixed with 100 mL of 20 mg/L of MG solution and agitated for 5 min for full adsorption, and then settled for 30 min. After adsorption the removal efficiency was found to be ~98%. Thereafter the desorption studies were carried out by adding the MG adsolubilized SMA (1 g) into 25 mL of acetone and the mixture was agitated for 10 min. After desorption, >90% of the MG dye could be recovered. At that moment no SDS was desorbed. Then the MG released SMA was separated smoothly and washed with double distilled water and finally dried at 333 K for further adsorption experiments. The regenerated SMA was employed again for the removal of MG, and the removal efficiency of SMA at that stage was ~80%. The organic solvent could be recovered on distillation, and reused. The recovered MG dye could also be recycled. This study reveals that SMA can be reused economically. Figure 7.7 illustrates the schematic flow diagram of recovery of MG and regeneration of SMA.

Alternatively, aqueous sodium hydroxide (0.25 M NaOH) solution could be used for desorbing MG and SDS simultaneously from the exhausted SMA to get back the pristine alumina and alkaline dye solution (Adak and Pal 2006; Das et al. 2009a).

Continuous Fixed-Bed Adsorption Column for Lab-Scale Experiment

Continuous fixed-bed adsorption column studies were carried out to examine the dynamic behaviour of MG dye removal by surfactant-loaded alumina (Das et al. 2009b). Continuous flow adsorption experiments were accomplished in borosilicate

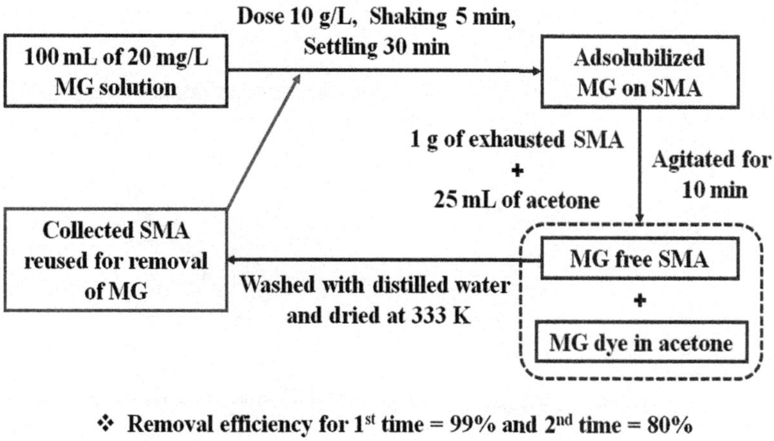

Fig. 7.7 Diagram showing the recovery of MG dye and regeneration of SMA

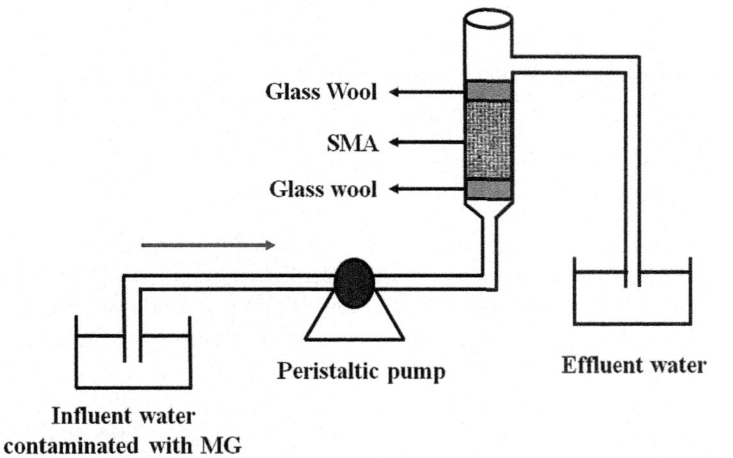

Fig. 7.8 Schematic diagram of fixed-bed adsorption column

glass columns of 2 cm internal diameter and 55 cm length. A schematic diagram of the experimental set-up is shown in Fig. 7.8. The column was packed with known quantity of SMA between two layers of glass wool to attain a particular bed height and to prevent loss of SMA during flow of MG solution inside the column, and also to ensure a closely packed arrangement. The MG dye solution was pumped upward to avoid channelling due to gravity. The MG bearing (concentration 100 mg/L) synthetic water was allowed upward through the bed layer at volumetric flow rate of 9 mL/min (~1.72 $m^3/m^2/h$). Here Logit model was applied to develop the column design parameters, and the bed depth selected for it was 10 cm. The experiment was conducted at room temperature, samples were collected at regular time intervals and were investigated for residual MG concentration. Experiment was also carried out

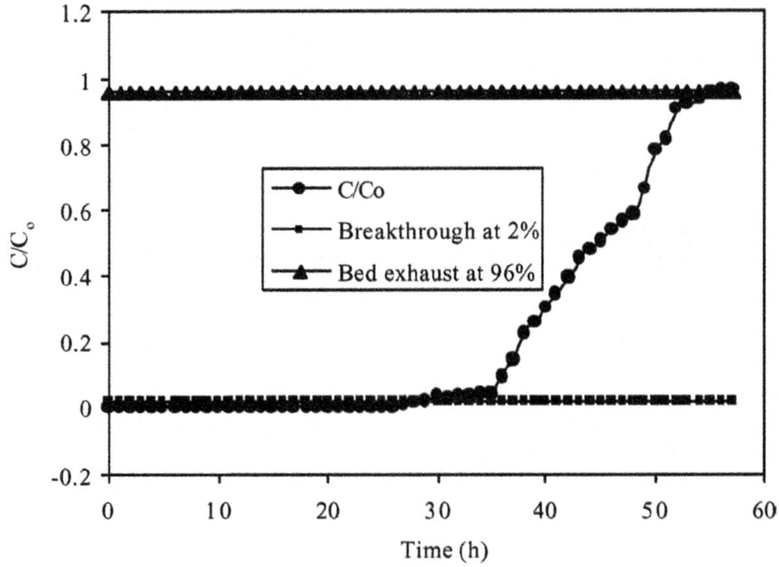

Fig. 7.9 Breakthrough curve for MG bearing synthetic water (Das et al. 2009b)

with real MG bearing wastewater (concentration 396.54 mg/L) keeping the bed depth as 10 cm and flow rate of 9 mL/min.

Under the specified condition, the fixed-bed column experiments were conducted and the observed results were utilized for constructing the breakthrough curves (C/C_0 vs. time) in order to evaluate the breakthrough time and exhaust time. In addition, the amount of MG bearing water treated was evaluated using the breakthrough time, exhaust time and flow rate. During the experiment, the height of adsorption zone and the rate at which it was moving through the bed were also evaluated. Fig. 7.9 shows the breakthrough curve for the adsorption of MG bearing synthetic water using SMA (Das et al. 2009b).

The fixed-bed column was designed by Logit method and could be written as Eq. 7.1,

$$\ln\left[\frac{C/C_0}{1-C/C_0}\right] = -\frac{KNX}{V} + KC_0t \tag{7.1}$$

where C_0 is the initial dye concentration (mg/L), C is the dye concentration at any time t (mg/L), V is the approach velocity (cm/hr), X is the bed depth (cm), K is the adsorption rate constant (L/(mg h)) and N is the adsorption capacity coefficient (mg/L).

The value of C corresponding to the ratio of C/C_0 could be calculated from Fig. 7.9. It can be utilized to find out the percentage of the total column saturated at breakthrough stage. The breakthrough time and exhaust time were found to be 29 h and 57 h respectively. At this stage, the corresponding volumes of MG bearing synthetic water were 15.66 and 30.78 L respectively. The rate at which the adsorption zone was moving through the bed was found to be 0.26 cm/h and the height of adsorption zone was 6.30 cm.

The equations for minimum bed depth (x_0) and the percentage of the total column saturated at breakthrough can be written as:

$$x_0 = \frac{V}{KN} \ln\left(\frac{C_0}{C_B} - 1\right) \tag{7.2}$$

$$\%\text{saturation} = \frac{h + (F - 1)h_z}{h} \tag{7.3}$$

$$F = \frac{S_z}{S_{max}} = \frac{\int_{V_B}^{V_E} (C_0 - C)dV}{C_0(V_E - V_B)} \tag{7.4}$$

where h is the total bed depth (cm), F is the fraction of solute present, S_z is the amount of dye that has been removed by the adsorption zone from breakthrough to exhaustion, S_{max} is the amount of solute removed by the adsorption zone if completely exhausted, h_z is the height of exchange zone (cm), V_E is the total volume of MG bearing synthetic water treated to the point of exhaustion (L), V_B is the total volume of MG bearing synthetic water treated to the point of breakthrough (L), x_0 is the minimum bed depth (cm) and C_B is the desired concentration of MG at breakthrough (mg/L).

Hence, the percentage of the total column saturated at breakthrough for the removal of MG bearing synthetic water using SMA was found to be ~71.72% and the minimum bed depth was 3.33 cm. These credentials can be employed for designing the adsorption column.

The adsorptive potential of SMA on real MG bearing wastewater was examined by fixed-bed column experiments. The real colorant wastewater was procured from a small scale handloom industry and the MG concentration in the wastewater was found to be 396.54 mg/L. The column study was further performed with MG bearing wastewater under precisely identical conditions. Here the breakthrough time was observed after 18 h (with breakthrough concentration of 2 mg/L, $C/C_0 = 0.0051$) and the exhaustion point (at 90%) achieved after 37 h of continuous flowing of dye wastewater (Fig. 7.10). The saturation percentage and minimum bed depth of the column were found to be 93.04% and 5.92 cm respectively. They also concluded that the column efficiency for the MG bearing wastewater was much better as compared to the MG bearing synthetic water.

The exhausted SMA from the fixed-bed column was regenerated by employing acetone as the eluting agent by allowing in an up-flow mode with a flow rate of 5 mL/min (~0.96 $m^3/m^2/h$) and was continued until the recovery of ~98% of the adsorbed MG. After that the bed was reconstructed by passing the double distilled water for 5 h in an identical direction with the flow rate of 5 mL/min. It was

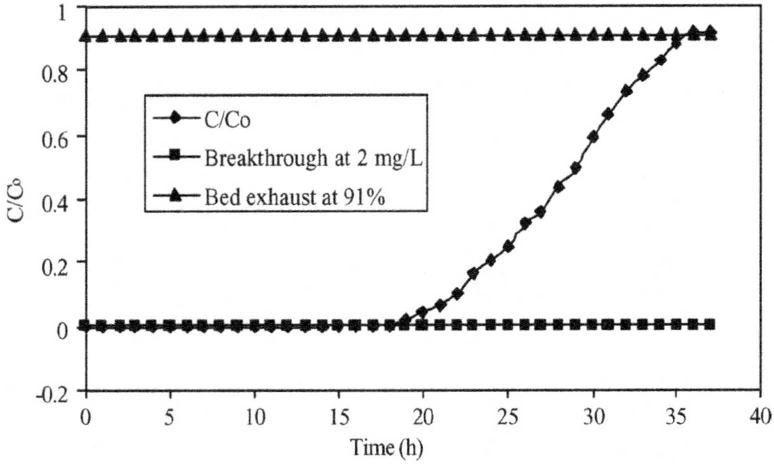

Fig. 7.10 Breakthrough curve for adsorption of MG bearing wastewater onto SMA (Das et al. 2009b)

concluded that the regenerated SMA reveals a lower efficiency compared to the pristine material (Das et al. 2009b).

5.2 *Removal of Mn(II), Ni(II) and Cu(II) Using SMA from Aqueous Environment*

Industrial wastewater containing toxic metals, predominantly manganese (Mn(II)), nickel (Ni(II)) and copper (Cu(II)) causes serious threat to the environment owing to the non-degradable characteristic of the heavy metals. The presence of such metals in various industrial applications such as mining, battery manufacturing, electroplating, metal finishing, etc. can have adverse effect on human health, animals and ecological systems if protective measures are not taken. The exposure to heavy metals beyond the permissible limits can cause heart disorders, liver, kidney and central nervous system damage, respiratory failure, and even death. Therefore, removal of heavy metals from wastewater is crucial and necessary.

Over the years, various conventional techniques including chemical precipitation, ion exchange, electrolyte recovery, membrane separation, flotation and adsorption have been proposed for the treatment of aforementioned heavy metals. However, among these techniques some involve high capital investment and operational cost, and some are not practicable for complete removal of heavy metals when present at low concentrations. Some literatures have proposed adsorption to be the most promising method for metal ion removal due to its environmental friendliness, simplicity of design, low cost, ease of operation without causing secondary pollution and the opportunity of being regenerated by incorporating the desorption methods.

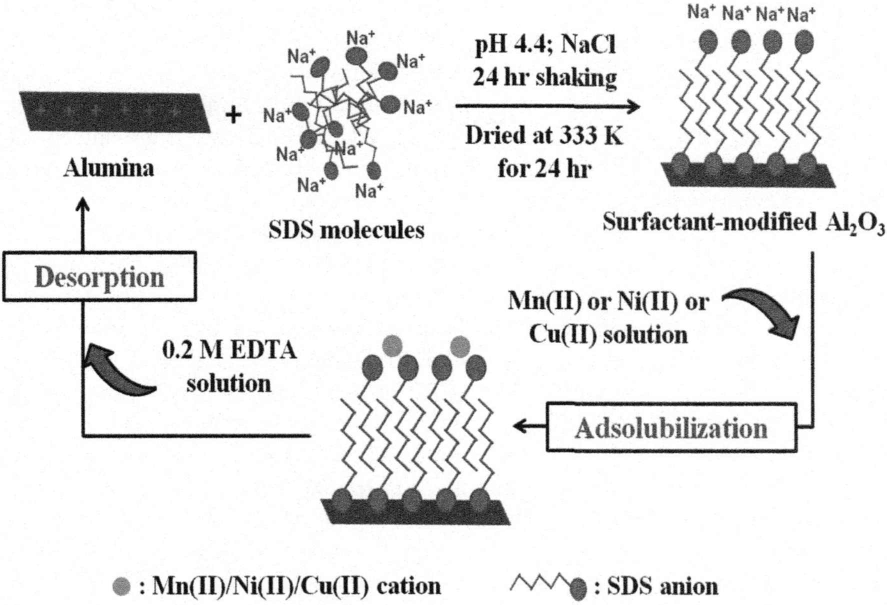

Fig. 7.11 Scheme of adsorptive behaviour of SMA on metal ions

In the recent years, surfactant treated solid surfaces such as modified alumina have been employed for adsolubilizing the organic pollutants (e.g. cationic dyes) (Adak et al. 2005; Das et al. 2009a, b). Under enhanced loading condition, the surfactant bilayer (also known as admicelle) were formed on the alumina surface, which had the capability to solubilize the organic molecules either by hydrophobic or electrostatic interaction. The same adsorbent material i.e. surfactant-modified alumina (SMA) has recently been used for the effective removal of heavy metals such as Mn(II), Ni(II) and Cu(II) from aqueous solution (Khobragade and Pal 2014, 2015). The surfactant-loaded alumina acquires the prospective advantages like easy-to-make chemical modification, easy regeneration and less disposal problem after adsorption. Most importantly, it also has the ability to attract the metal ions very fast. The proposed scheme for modification of alumina surface and adsolubilization of metal ions on SMA is presented in Fig. 7.11.

The batch experiments were carried out for the removal of Mn(II), Ni(II) and Cu (II) ions at different pH conditions because most of the heavy metals are precipitated at higher pH. In adsorption studies the samples were taken in glass vials of known capacity, sealed with stoppers and the required quantity of dried SMA was weighed and added to it. In this regard 20 mL of synthetic sample containing Mn(II), Ni(II) or Cu(II) with an initial concentration of 20 mg/L was added separately to the glass vials. The samples were agitated for a predefined contact period in an orbital shaking incubator at 150 rpm and at 303 K. After this the samples were filtered and the supernatant was collected. The absorbance of each filtrate was examined at a maximum wavelength of 279.5 nm for Mn(II), 232.0 nm for Ni(II) and 324.8 nm for Cu(II) using flame atomic absorption spectroscopy.

The optimum adsorptive conditions for the removal of Mn(II) ion were found to be pH 6.2, SMA dose 20 g/L and contact time 30 min. Similarly, the optimum conditions for the removal of Ni(II) were observed as pH 8.3, dose 5 g/L and contact time 60 min, whereas the optimum conditions for Cu(II) removal were computed as pH 5.3, adsorbent dose 4 g/L and contact time 75 min (Khobragade et al. 2016).

Adsorption kinetics are useful to demonstrate the sorption mechanisms for batch experimental study. The adsorption of metal ions onto surfactant-loaded alumina reached equilibrium in a shorter contact time resulting in fast and efficient metal removal, which would be beneficial for practical purposes. It was scrutinized that the pseudo-second-order kinetic model was best followed in most cases. Adsorption of metal ions on SMA pursued the Freundlich isotherm and the adsorption capacity demonstrated from Langmuir isotherm was found to be 1.22 mg/g for Mn(II), 6.87 mg/g for Ni(II) and 9.34 mg/g for Cu(II). The thermodynamic parameters were analyzed and it explored that the adsorption processes were favourable, spontaneous and endothermic. The applicability of SMA was further examined with metal-bearing wastewater collected from an electroplating industry. In addition, the adsorbed metal ions could be desorbed efficiently from the exhausted SMA by employing 0.2 M Na_2-EDTA (i.e. Disodium ethylenediaminetetraacetic acid) solution. It was examined that up to ~91% and ~92% recovery was plausible for Ni (II) and Cu(II) from exhausted SMA in 1 hr. upon treatment with it. On the other hand, ~92% desorption was achieved within 30 min in case of Mn(II). The regenerated SMA after Na_2-EDTA treatment was further utilized for Mn(II), Ni (II) or Cu(II) adsorption but the removal efficacy was diminishing significantly. Furthermore, Das et al. (2009a) have already demonstrated that aqueous NaOH solution (0.25 M) could desorb SDS from SMA. The results reveal that the exhausted SMA after treatment with 0.25 M NaOH followed by thorough wash with tap water and double distilled water brought back the alumina, which could be treated again with SDS solution (concentration 20,000 mg/L) to prepare SMA using the same procedure as described earlier in Fig. 7.4.

In addition, the fixed-bed column experiments were performed using a glass column having an internal diameter of 1 cm and height of 30 cm (Khobragade and Pal 2014). The column was packed with the adsorbent (i.e. SMA) equal to the required depth between two layers of glass wool. The column was pumped with metal-bearing synthetic water (Mn(II)/Ni(II)/Cu(II) concentration 20 mg/L) in an up-flow mode at a flow rate of 7 mL/min (~5.34 $m^3/m^2/h$) by employing a peristaltic pump. The samples were extracted at regular intervals of time and the residual heavy metal concentrations were analyzed in AAS. Logit method was applied to evaluate the column design parameters. The values of breakthrough time, exhaust time, amount of wastewater treated and height of adsorption zone for different metal ions are illustrated in Table 7.1. Based on aforesaid observations, the percentage of total column saturated at breakthrough for the removal of Mn(II), Ni(II) and Cu (II) ions using SMA were found to be 94.09% (Khobragade and Pal 2014), ~95% (Khobragade and Pal 2016) and 96.35% (Khobragade et al. 2017), respectively.

Table 7.1 Parameters for the column adsorption of Mn(II), Ni(II) and Cu(II) from aqueous solution using SMA

	Heavy metal ions		
Parameter	Mn(II)	Ni(II)	Cu(II)
Breakthrough time corresponding to C/C_0	0.58 hr ($C/C_0 = 0.091$)	2.67 hr ($C/C_0 = 0.089$)	5.5 hr ($C/C_0 = 0.092$)
Exhaust time corresponding to C/C_0	1.5 hr ($C/C_0 = 0.902$)	5.0 hr ($C/C_0 = 0.919$)	9.0 hr ($C/C_0 = 0.913$)
Treated volume at breakthrough time	0.245 L	1.12 L	2.31 L
Treated volume at exhaust time	0.630 L	2.10 L	3.78 L
Height of adsorption zone	6.49 cm	4.86 cm	4.03 cm
Rate at which adsorption zone was moving through the bed	7.06 cm/hr	2.08 cm/hr	1.07 cm/hr

5.3 Application of Surfactant-Modified Alumina (SMA) to Synthesize a Fenton-Type of Catalyst

During the past decades, the advanced oxidation processes has gained popularity for the effective treatment of recalcitrant organic pollutants. These methods produce highly reactive and non-selective hydroxyl ($^{\cdot}OH$) radicals which accelerate the oxidation and degradation of the compound until the attainment of complete mineralization of the target pollutants. Traditionally, Fenton-type reactions are employed for complete mineralization of pollutants and it has immense applications for the treatment of industrial wastewater. However, from practical point of view there are major constraints for traditional Fenton reaction which involves requirement of stringent acidic condition (pH 2.5–3.5), huge sludge production, non-recylability of the iron catalyst, etc. To triumph over these disadvantages, various types of heterogeneous catalysts have been synthesized to degrade recalcitrant organic pollutants through Fenton-like reactions (Bokare and Choi, 2014; Mahamallik and Pal 2016). These modified catalysts can usually execute over a wide range of pH 3–10. In these cases the oxidation is very slow compared to heterogeneous Fenton processes. In many cases the heterogeneous reaction shows higher efficiency compared to homogeneous systems as it consumes a low amount of hydrogen peroxide per mole of pollutant (Ahuja, 2014).

Heterogeneous Fenton reaction consists of two steps, namely adsorption and degradation. Nevertheless, the chances and extent of adsorption in most cases and its influence on degradation were overlooked. In recent times, Mahamallik and Pal (2016) have employed the adsolubilization theory in photo-Fenton type of reaction. They proposed that the adsolubilization concept could be utilized for the fabrication of heterogeneous material in order to increase their catalytic activity for the mineralization process. In this study, the cobalt (Co(II)) ion was attached on the SMA surface through adsolubilization by referring the identical procedure as discussed earlier in Fig. 7.11. The Co(II) loaded surfactant-modified alumina (designated as Co-SMA) has been considered as heterogeneous surface for conveying the Fenton

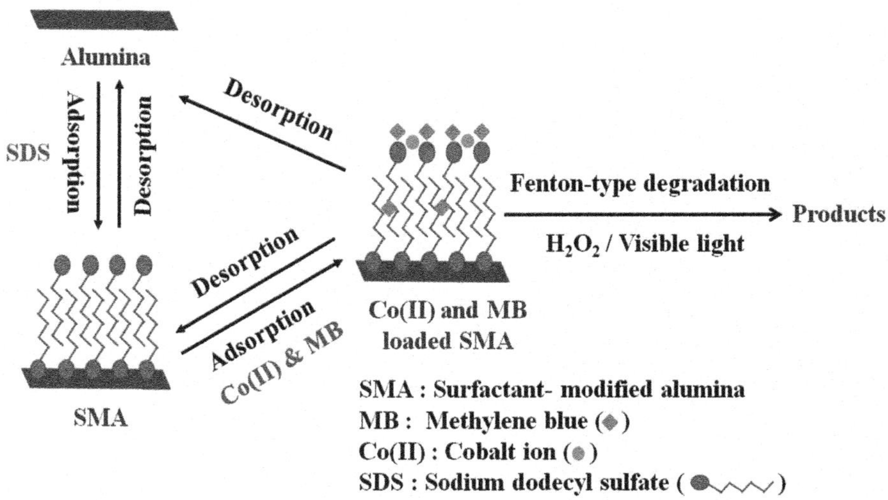

Fig. 7.12 Pictorial representation for the presentation of Co-SMA and its application in MB degradation

process. The obtained heterogeneous surface possesses the ability to accommodate the cationic dye (e.g. methylene blue (MB)) effectively (Fig. 7.12). They also reported that the photo-assisted Fenton reaction is helpful in accelerating the reaction. It is because the photon helps in reducing the metal ions e.g. Co(III) to Co(II). The uniqueness of this study is that the degradation of MB entirely occurs on a solid surface and the monitoring of the reaction has been done through an extraction technique. It was also noticed that the obtained catalyst could be recycled up to third cycle without loosing the efficiency significantly.

In another study, the Co-SMA was further employed for the degradation of methyl orange (MO) in presence of H_2O_2 and visible light irradiation under heterogeneous photo-Fenton process (Mahamallik and Pal 2017). Also, the degradation of textile wastewater was demonstrated under similar operational conditions. Adsolubilization concept was also used in 4-nitrophenol reduction using borohydride as a reducing agent and gold nanoparticles as catalyst (Mahamallik and Pal 2015). The incorporation of the catalyst and 4-nitrophenol together on the solid substrate makes the reaction more feasible.

6 Conclusions

We describe here how alumina, an abundantly occurring cheap and non-toxic material can remove anionic surfactant such as sodium dodecyl sulfate (SDS), when present in water at high concentration. After the removal of SDS, the derived surfactant-modified alumina (SMA) can be exploited as a potential material in terms of pollutant (such as dye) recovery and reuse. The discovery of SMA is stupendous in the perception that SDS can form a bilayer structure called admicelle on alumina surface under controlled conditions. This surfactant bilayer possesses the ability to solubilize the cationic dyes (namely, malachite green (MG), crystal violet (CV), methylene blue (MB)) as well as heavy metal ions (namely, Mn(II), Ni(II), Cu(II), Co(II)), either by electrostatic or hydrophobic interactions. This adsorption process is very fast and the phenomenon is recognized as 'adsolubilization'. It is easy to get back the pollutants as well as the SMA/alumina after the removal of target pollutants by employing low-cost treatment and then reused for subsequent experiments. Herein the dye or metal ion can be recycled and hence the 'zero-waste discharge' is perceived. Moreover, the SMA sludge obtained after the adsorptive removal of metal ion (especially Co(II)) can be used as a heterogeneous catalyst for Fenton-type reactions in order to degrade the hazardous dyes. The cobalt-loaded SMA (Co-SMA) is thus utilized further as a catalyst for the degradation of MB dye through a heterogeneous photo-Fenton process. Hence the 'waste to wealth' practice is realized.

The interesting aspect of the aforementioned study is that each intermediate product is utilized in a fruitful way without any loss, which is the principle of 'closed-loop system' in industrial process. In addition, all the materials utilized here are non-toxic, commercially obtainable and inexpensive. This technique leads to environment-friendly 'zero-waste concept' and fulfils the prime focus of '*Green Chemistry*' approach. By exploitation of organized chemical procedures, we are able to diminish the waste of materials, sustain the 100% atom economy and forestall the use of dangerous chemicals. Overall, the process can lead to a sustainable approach for minimization and prevention of pollutants.

Acknowledgement We are extremely grateful to Dr. Prateeksha Mahamallik for her helpful contributions. The book chapter is much improved as a consequence.

References

Adak, A. and Pal, A. (2006). Removal of phenol from aquatic environment by SDS-modified alumina: Batch and fixed bed studies. *Separation and Purification Technology*, **50(2):** 256–262. doi:https://doi.org/10.1016/j.seppur.2005.11.033.

Adak, A., Bandyopadhyay, M. and Pal, A. (2005). Adsorption of anionic surfactant on alumina and reuse of the surfactant-modified alumina for the removal of crystal violet from aquatic environment. *Journal of Environmental Science and Health, Part A*, **40(1):** 167–182. doi:https://doi.org/10.1081/ESE-200038392.

Ahuja, S. (2014). Overview: Water Reclamation and Sustainability. *In:* Water Reclamation and Sustainability. Elsevier, Boston, 1–18.

Anastas, P.T. and Warner, J.C. (1998). Green Chemistry: Theory and Practice. Oxford University Press, New York.

Ayranci, E. and Duman, O. (2007). Removal of anionic surfactants from aqueous solutions by adsorption onto high area activated carbon cloth studied by in situ UV spectroscopy. *Journal of Hazardous Materials*, **148(1–2):** 75–82. doi:https://doi.org/10.1016/j.jhazmat.2007.02.006.

Bokare, A.D. and Choi, W. (2014). Review of iron-free Fenton-like systems for activating H_2O_2 in advanced oxidation processes. *Journal of Hazardous Materials*, **275:** 121–135. doi:https://doi.org/10.1016/j.jhazmat.2014.04.054.

Das, A., Saha, S., Pal, A. and Maji, S.K. (2009a). Surfactant-modified alumina: An efficient adsorbent for malachite green removal from water environment. *Journal of Environmental Science and Health, Part A*, **44(9):** 896–905. doi:https://doi.org/10.1080/10934520902958708.

Das, A., Pal, A., Saha, S. and Maji, S.K. (2009b). Behaviour of fixed-bed column for the adsorption of malachite green on surfactant-modified alumina. *Journal of Environmental Science and Health, Part A*, **44(3):** 265–272. doi:https://doi.org/10.1080/10934520802597929.

Draths, K.M. and Frost, J.W. (1994). Environmentally compatible synthesis of adipic acid from D-glucose. *Journal of the American Chemical Society*, **116(1):** 399–400. doi:https://doi.org/10.1021/ja00080a057.

Khobragade, M.U. and Pal, A. (2014). Investigation on the adsorption of Mn(II) on surfactant-modified alumina: Batch and column studies. *Journal of Environmental Chemical Engineering*, **2(4):** 2295–2305. doi:https://doi.org/10.1016/j.jece.2014.10.008.

Khobragade, M.U. and Pal, A. (2015). Adsorptive removal of Cu(II) and Ni(II) from single-metal, binary-metal, and industrial wastewater systems by surfactant-modified alumina. *Journal of Environmental Science and Health, Part A*, **50(4):** 385–395. doi:https://doi.org/10.1080/10934529.2015.987535.

Khobragade, M.U. and Pal, A. (2016). Batch and continuous fixed-bed column adsorption for the removal of Ni(II) from aqueous solutions using surfactant-treatment alumina. *Recent Patents on Engineering*, **10(1):** 36–50. doi:https://doi.org/10.2174/1872212110666151218204647.

Khobragade, M.U., Nayak, A.K. and Pal, A. (2016). Application of response surface methodology to evaluate the removal efficiency of Mn(II), Ni(II), and Cu(II) by surfactant-modified alumina. *Clean Technologies and Environmental Policy*, **18(4):** 1003–1020. doi:https://doi.org/10.1007/s10098-016-1116-0.

Khobragade, M.U., Nayak, A.K. and Pal, A. (2017). Solid-phase extraction of Cu(II) from aqueous solution using surfactant-modified alumina. *Journal of Hazardous, Toxic, and Radioactive Waste*, **21(2):** 04016017. doi:https://doi.org/10.1061/(ASCE)HZ.2153-5515.0000336.

Mahamallik, P. and Pal, A. (2015). A soft-template mediated approach for Au(0) formation on a heterosilica surface and synergism in the catalytic reduction of 4-nitrophenol. *RSC Advances*, **5:** 78006–78016. doi:https://doi.org/10.1039/C5RA16655A.

Mahamallik, P. and Pal, A. (2016). Photo-Fenton process in a Co(II)-adsorbed micellar soft-template on an alumina support for rapid methylene blue degradation. *RSC Advances*, **6:** 100876–100890. doi:https://doi.org/10.1039/C6RA19857K.

Mahamallik, P. and Pal, A. (2017). Degradation of textile wastewater by modified photo-Fenton process: Application of Co(II) adsorbed surfactant-modified alumina as heterogeneous catalyst. *Journal of Environmental Chemical Engineering*, **5(3):** 2886–2893. doi:https://doi.org/10.1016/j.jece.2017.05.044.

Schouten, N., van der Ham, L.G.J., Euverink, G.-J.W. and de Haan, A.B. (2007). Selection and evaluation of adsorbents for the removal of anionic surfactants from laundry rinsing water. *Water Research*, **41(18):** 4233–4241. doi:https://doi.org/10.1016/j.watres.2007.05.044.

Chapter 8
An Integrated Solid Waste Management (ISWM) Plan Using Google Earth and Linear Programming: A Case Study of Kharagpur City, West Bengal

Abhishek Singhal and Sudha Goel

1 Introduction

Solid Waste Management (SWM) is an obligatory function of the urban local bodies (ULBs) such as a municipality. Today, ULBs in developing countries like India are creating new infrastructure for better SWM and are trying to change old and unsustainable practices. A major issue in current SWM practices in developing nations is availability of timely data and financial resources. Remote Sensing (RS) and GIS can be used together to develop an integrated solid waste management plan since RS provides spatial and temporal data readily in the form of satellite images available on the Internet and GIS allows mapping with overlays of population distribution, for locating bins, vehicle routing and facility siting. A major advantage of using RS and GIS is that they can provide accurate, inexpensive, wide-ranging and rapid results which can be further combined with other tools like linear programming. This approach has been used for various applications in several cities around the world and was reviewed by Dutta and Goel (2017).

The objective of this study was to evaluate deficiencies in current solid waste management practices in Kharagpur city and propose an integrated solid waste management plan for a cleaner city using novel methods like remote sensing (RS) and Geographic Information Systems (GIS) along with linear programming for optimal vehicle routing.

A. Singhal (✉) · S. Goel
Civil Engineering Department, Indian Institute of Technology Kharagpur, Kharagpur, West Bengal, India

D. Sengupta et al. (eds.), *Treatment and Disposal of Solid and Hazardous Wastes*,
https://doi.org/10.1007/978-3-030-29643-8_8

2 Description of Study Area – Kharagpur

Kharagpur is an important city in Paschim Medinipur district of West Bengal and was established as one of India's major rail junctions in 1898. Kharagpur is located at Latitude 22.330239 °N and Longitude 87.323653 °E covering an area of about 91 km^2. It has an average elevation of 29 m (95 ft.) above Mean Sea Level.

The Kharagpur urban area consists of the Kharagpur municipal area as well as railway settlements which are not included in the jurisdiction of Kharagpur municipality (Fig. 8.1). Kharagpur municipal area has a population of 289,129 (Census of India 2011) spread over 91 km^2 area and is divided into 35 wards (Kharagpur Municipality, 2016). Based on Census data and assuming exponential growth, the population of Kharagpur Municipality in 2016 was estimated to be 314,628.

The campus of Indian Institute of Technology (IIT) Kharagpur is designated as Ward No. 30 and is inside the municipal area. The Indian Air Force has two airbases close to Kharagpur: Kalaikunda and Salua.

2.1 Generation and Characterization of Waste

Major sources of MSW in the Kharagpur area are residential areas, commercial/market areas, government offices, hotels/restaurants, fruit and vegetable markets, and institutions. Kharagpur Municipality is responsible for the collection, treatment, and disposal of solid waste according to Indian regulations. Based on the current estimated population of Kharagpur of 314,628 and assuming a per capita waste generation rate of 0.5 kg/capita-day, the total amount of solid waste generated in Kharagpur is estimated to be 158 metric tons (MT) d^{-1}.

MSW from Kharagpur was characterized and recyclables such as plastic, paper, metals and textiles constituted 19.6% of the total MSW generated in Kharagpur (Kumar and Goel 2009). The remaining fraction (80.4%) was mixed residue, i.e., organic material mixed with soil, sand, mud and other inert materials. Mixed residue had an organic carbon content of 8.92 (±5.92) %, volatile solid content of 19.63 (±9.53) % and fixed solids content of 80.35 (± 9.54) %. These results show the high inert content of the mixed waste which is mostly non-toxic and benign in nature. The moisture content of the waste was 42.05 (±10.05) %.

2.2 Collection and Storage of MSW

Waste from residential and commercial establishments is generally disposed of by residents in nearby community bins. The existing bins are stationary enclosures (also called vats) made up of plain concrete (Fig. 8.2). The sizes of the bins at every location are not the same, and vary depending on site and roadside conditions.

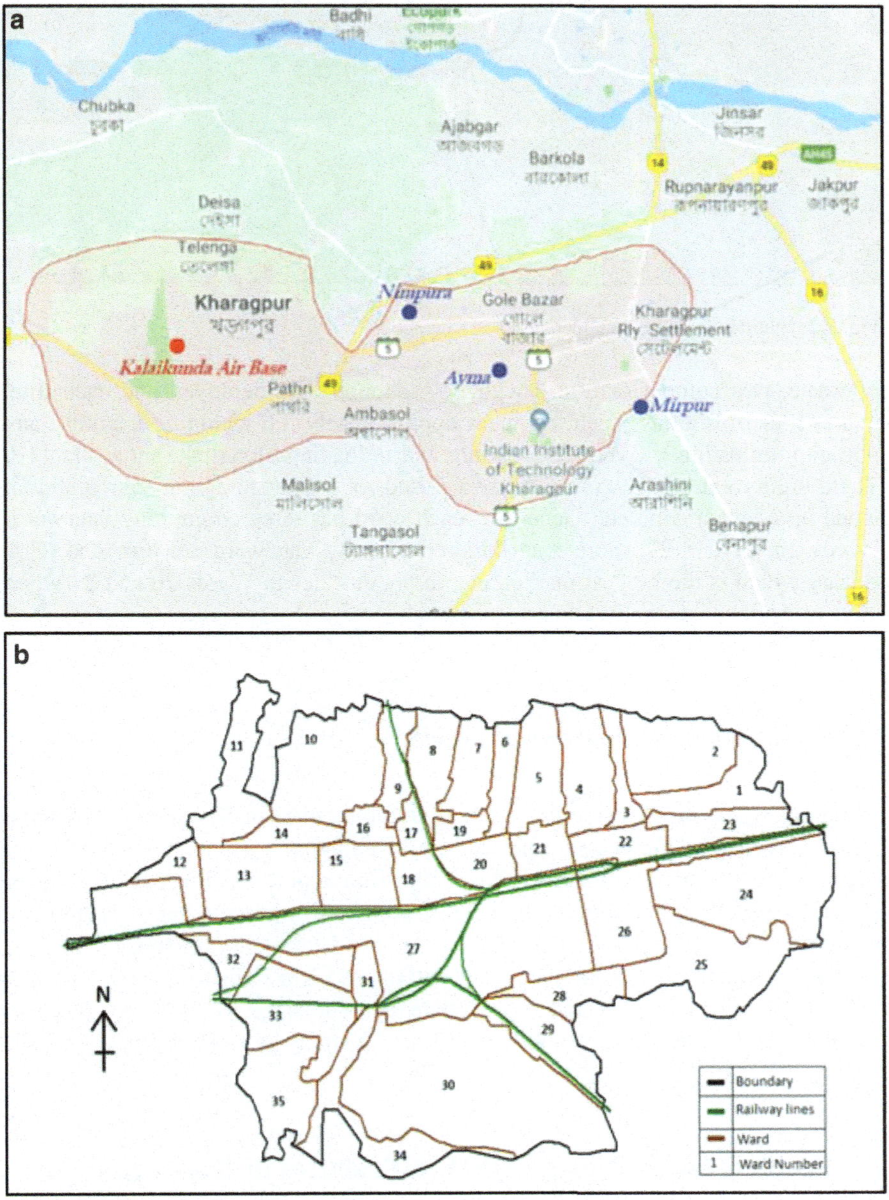

Fig. 8.1 (**a**) Map of Kharagpur with dump sites and air force base locations and (**b**) Ward-wise map of Kharagpur

Collection of waste is done by the Kharagpur municipality after every 3–4 days. Two compactor trucks and five tractor-trolleys are used to collect the waste generated in the city. Sometimes large amounts of waste are piled by the road-side and are collected using a JCB backhoe loader and loaded into a tractor-trolley. For collection

Fig. 8.2 Stationary waste containers used in Kharagpur

of waste, two compactor trucks with 5–7 labourers are deployed and each truck makes four trips a day (each trip takes approximately 2 h including loading-transportation-unloading). Generally, loading and unloading alone take more than 1 h. There are a total of 112 community vats used for waste storage at source and are made up of plain concrete. Generally, each ward has three community vats but in Wards 20, 21 and 27, more than three community vats/ward are provided. Gole Bazaar which is the biggest market area in the city lies in Wards 20 and 21 where huge amounts of market waste are generated.

2.3 Treatment and Disposal of MSW

Nimpura, Ayma and Mirpur are the three dumping sites in Kharagpur. Municipal waste is regularly dumped on open land by Kharagpur municipality. There is no MSW treatment facility or sanitary landfill in Kharagpur; all three sites mentioned above are just open dumping grounds. Soil from the Mirpur site has a considerable amount of clay and both Nimpura and Ayma have laterite morum soil.

All three sites mentioned above are within a 15 km radius of the Kalaikunda Airbase of Indian Air Force and as per Indian regulations, i.e., SWM Rules 2016, the location of any disposal site should be at least 20 km away from the runway.

3 Problems with Present MSWM Practices in Kharagpur

3.1 Source Segregation and Waste Treatment

Ideally, waste should be segregated at source into three fractions:

(i) biodegradable waste like food and yard waste,
(ii) recyclables like paper, plastics, metals and glass, and
(iii) mixed waste or non-biodegradable waste.

However, very few areas in Kharagpur have source segregation of waste and the collection trucks are not designed for the collection of segregated waste which requires more than one compartment. Generally, all waste collected by the Municipality is unsegregated and is crudely dumped on open ground without any treatment. The absence of source segregation makes the treatment of waste very difficult and expensive.

3.2 Underestimation of Per Capita Waste Generation Rates

Kharagpur Municipality had estimated the average per capita generation rate to be 350 g/capita-day in 2007 (Kharagpur Municipality). They continue to use this estimate for designing their collection systems resulting in underestimation of collection equipment (containers and vehicles) requirements and inadequate collection (around half of the waste generated in Kharagpur is left uncollected daily). Uncollected waste results in health and odour problems, choking of storm water drains, and overflowing dustbins.

3.3 Collection

There are two major issues with waste collection in Kharagpur:

Bin location and design: Door-to-door collection is not provided in Kharagpur as is the case in many other cities like Kolkata, Ahmedabad, Mysuru, Guwahati and Surat. Waste from residential and commercial establishments are disposed by residents in nearby community bins. Current locations of the bins are not uniform, some are too far from houses and shops, and residents are often unwilling to walk a long distance in search of a bin. Further, waste is collected once every 3–4 days but bins are not designed for holding waste for this period of time. Due to all these reasons, illegal dumping of waste in the local area is common (Fig. 8.3). Stray animals are attracted to the scattered waste resulting in safety issues for residents. They also prevent municipal workers from attending to the bins and create traffic problems.

The existing bins are stationary enclosures made of plain concrete. These concrete walls break easily and currently, many of the community bins are in broken condition.

Vehicles: Kharagpur Municipality does not have enough vehicles to collect all the waste generated in the city. The present vehicles used for collection are garbage compactor trucks, mini-trucks, and tractor-trolleys. Compactor trucks are adequate for waste collection but simple cargo trucks and tractor-trolleys are not fit

Fig. 8.3 Illegal dumping ground near IIT Kgp flyover (**a**. Dumping site, **b**. Google Earth image of the site)

for collection since they scatter waste and drip leachate (liquid coming out of biodegradable waste) while transporting waste, more so during the monsoon season. Also, most of the vehicles in current use are more than 5–10 years old and many are in deplorable condition.

3.4 Lack of Financial and Human Resources

Currently, Kharagpur municipality does not have sufficient financial and human resources to collect all the waste generated in the city. Numbers and capacities of collection vehicles and sanitary workers for waste collection are highly inadequate. The recommended benchmark for sanitary workers is 2.8 workers/1000 population (Gupta et al. 1998). An estimated 300 persons were employed as labourers in 2016 for solid waste management in Kharagpur. For the estimated population of 314,628 persons in 2016, this amounts to 0.95 workers/1000 people which is well below the benchmark. In general, municipal workers have high rates of absenteeism and low productivity, partly due to public attitude and habits. Due to lack of financial resources, the Municipality does not have enough safety equipment for workers resulting in injuries. Due to manual sweeping and use of open carts, workers are exposed to waste causing health problems among the workers (Annepu 2012). Common health problems include respiratory and skin diseases, and infections in lungs or bladder.

3.5 Dumping Sites

There is no sanitary landfill in Kharagpur and all three sites mentioned previously are open dumping grounds. Sites in Mirpur and Nimpura are located close to densely populated residential areas and are responsible for causing health and foul smell-related problems in the nearby areas. Nimpura site is located about 30 m away from a school and is very close to NH-6. Ayma is relatively better located compared to other dumpsites (nearest settlement is 500 m away from the site). According to Indian regulations, a landfill should be located at least 500 m away from any residential settlement and State/National Highway (SWM Rules 2016). Both Nimpura and Mirpur are in contravention of these rules.

All three dumpsites are located in the proximity of the Kalaikunda Airbase of the Indian Air Force and as per Indian regulations, the location of any disposal site should be at least 20 km away from the runway (SWM Rules 2016). The disposal site at Nimpura is less than 5 km away, whereas the site in Ayma is around 8 km away. The other site at Mirpur is around 13–15 km away from the runway. There are many birds flying over these dumping sites due to which many accidents have occurred due to scavenger bird hits during take-off and landing of aircrafts.

Soil from the Mirpur site has considerable amounts of fine clay which has low permeability. Hence, ground water in this area is less vulnerable to contamination due to percolation of leachate generated from the solid waste dumped at these sites. On the other hand, Nimpura as well as Ayma have laterite morum soil, which is highly permeable, thus increasing the chance of groundwater contamination due to percolation of leachate at these sites.

4 Design and Planning of SWM System for Kharagpur

An integrated MSWM plan was developed for Kharagpur using Remote Sensing and Geographical Information System (GIS). Linear programming and all available data is shown in a flowchart (Fig. 8.4).

- First, all available data relevant to the preparation of an ISWM were collected. This included preparing a digitized ward map of Kharagpur and obtaining ward-

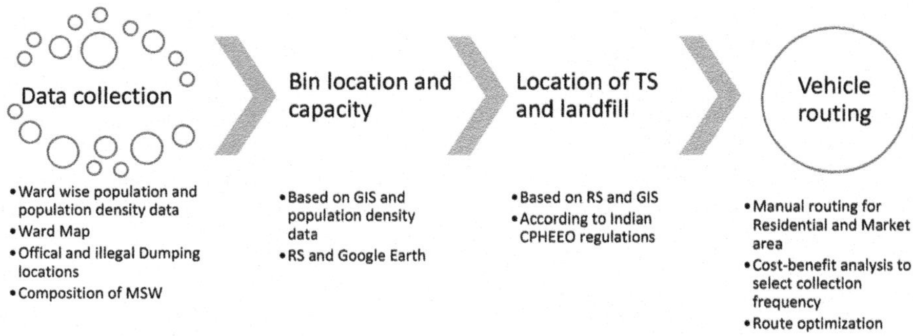

Fig. 8.4 Step-wise flowchart for developing an integrated ISWM plan for Kharagpur

wise population data. Locations of dumping sites and data regarding composition of MSW in Kharagpur were obtained from the literature. Known dumping sites were identified in satellite images and verified in a ground survey. Data regarding vehicles and labour employed were obtained from Kharagpur municipality officials.

- Second, bin location plans for residential and market area bins were developed for wards 20 and 21. Bin sizes for each area were calculated based on population density data.
- Third, siting of a transfer station and sanitary landfill was done using RS and GIS.
- Fourth, routing of garbage collection trucks in residential and market areas was optimized using linear programming and the results compared to manually drawn vehicle routes.

Google Earth Pro software and maps were used for developing the integrated solid waste management plan for Kharagpur.

4.1 Bin Location Plan and Bin Size Calculation

To solve problems related to overflowing waste bins, illegal dumping along roads and near houses and to improve the efficiency of the collection system in terms of time and fuel, adequate numbers of bins of proper size and location are required. Bin plans were developed for wards 20 and 21.

4.1.1 Methodology Used for Calculating Bin Size and Planning Bin Locations

Bin sizes and locations were determined using data from the literature and all assumptions are noted here:

- The bin size required depends on the density of MSW and the amount of waste generated alongwith a factor of safety. The density of the municipal solid waste generated in Kharagpur ranged from 250 to 500 kg/m^3 (Kharagpur Municipality, 2017). For an efficiently designed system, a per capita generation rate at the higher end and a low bulk density should be assumed (Goel 2008). So, for safety and greater efficiency even in extreme cases, 250 kg/m^3 was assumed as the density of MSW for design purposes.
- For source segregation, two bins are provided at each bin location. Both bins are placed close to each other and counted as a single bin location. One is "Mixed Waste" for storage of mixed waste and other is for "Recyclables" storage. This source segregation is applicable to both market and residential areas.
- The sizes of the 'Mixed waste' and 'Recyclables' bins in the residential areas are based on the composition of the waste generated in Kharagpur (Kumar and Goel 2009). MSW in Kharagpur has about 20% recyclables, so the size of the Recyclables bin should be 20% of the volume of the total waste brought to the bin location by the residents. The remaining 80% of the total waste, i.e., mixed residue, has to be deposited in the 'Mixed waste' bin. Therefore, the size of the mixed waste bin container should be 80% of the total waste. By using a factor of safety, additional space can be provided in the bins, and problems of overfilled bins and scattering of waste can be minimized.
- No data were available regarding the composition of waste generated in the market area in Kharagpur. Based on an on-going survey of MSW in Kolkata and Guwahati, two bins of equal capacity, i.e., 50% of the total weight of waste, were provided for the market areas.
- Based on Indian government guidelines, a bin has to be provided every 100 m ($\pm$ 25) along the road (Swachh Bharat Mission 2016).
- For design purposes, the per capita waste generation rate was assumed to be 500 g/capita-day.
- Based on Census of India, 2011, the number of people per household was 5 and was used in these calculations.
- A factor of safety (FOS) of 2 was provided to avoid overloading of bins.
- Daily collection of solid waste was provided due to the high moisture content of MSW, considerable amounts of biodegradable materials in the waste and extremely hot and humid climate.
- Bins in DIG and Sub-divisional offices are assumed to be the same size as market area bins due to similar composition of the waste and are collected along with the market waste. Bin sizes provided for them are 0.5 m^3 and 1 m^3 for DIG office and Sub-divisional office, respectively.

4.1.2 Bin Location Plan

Bin location plans for ward 20 are shown in Fig. 8.5. Each node/pin represents a bin location and consists of two bins: one for 'Mixed waste' and another for 'Recyclable' waste. Green coloured nodes are for bins located in residential areas. Red and

Fig. 8.5 Bin plan for Ward 20

brown nodes with green labels are for bins in apartments. Yellow nodes represent bins in market areas. Special bins are provided with red colour node and a red label for specific locations like temples, government offices, and schools that are different from market and residential areas. Bins for temples and schools are of the same size as bins in residential areas of that ward.

4.1.3 Calculation of Size of Bins in Residential Areas

In residential areas, sizes of the bins were designed on the basis of the number of residents living within a 100 m radius of the bin location. To calculate the number of people living within 100 m radius of a bin, population and population density data for each ward was used along with GIS. Google Earth software was used to calculate the ward area and empty area in the ward (like playgrounds, open lands, and children parks). An example of a bin size calculation for a residential bin is shown below; a factor of safety of 2 was assumed in all calculations.

Bin size calculation for Ward 20 *(represented by green pin on bin plan)*
Population of Ward 20 = 8543
Area of ward = 879,368 m^2
Empty area = 152,000 m^2
Populated area = 727,368 m^2
Area of circle with 100 m radius = 31,400 m^2/bin

(continued)

Population per 100 m radius circle = (8543/727,368) × 31,400 = 370 persons/bin (approximately)
Total waste generated = 370 × 0.5 kg/person-day = 185 kg/day
Total volume of waste, assuming bulk density of waste is 250 kg/m^3 = 185 (kg/d)/250 (kg/m^3) = 0.74 m^3/d
Size of mixed waste bin = 0.74 × 0.8 × 2 (FOS∗) =1.184 → 1.25 m^3 **
Size of recyclables bin = 0.74 × 0.2 × 2 (FOS∗) = 0.296 → 0.5 m^3 **
*FOS is a factor of safety for the bin
∗∗The final size of bin will be the larger available bin size

Therefore for Ward 20, the bin size is 1.25 m^3 for mixed waste and 0.5 m^3 for non-biodegradable waste bin. For Ward 21, the sizes required are 1.6 m^3 for mixed waste and 0.5 m^3 for non-biodegradable waste, respectively. Bins greater than 100 L capacity are too large and too heavy compared to common household bins. So, mechanical lifting is essential for emptying large bins into the garbage collection truck. Large bins can be substituted by two or more smaller bins.

There are some apartment buildings in Ward 20. Apartment bins have different sizes compared to other bins in the ward. There are a total of 20 apartment buildings having six floors and 20 buildings having three floors. Bin sizes for mixed waste in 3-storey and 6-storey apartment buildings were estimated using the same method and found to be 0.6 m^3 and 1.0 m^3 while sizes for recyclables bins were the minimum available size of 0.25 m^3 in both types of buildings.

4.1.4 Calculation of Size of Bins in Market Area

The methodology for designing bins for the market area is different from a residential area because the amount and composition of waste generated in the market are different from waste generated in residential areas. About one-third of the total waste generated in the city is commercial waste, i.e., 52.4 MT/day. Gole Bazar is the main market of Kharagpur city and it contributes about half of the total market waste generated in the city. Gole Bazar is located along the boundary dividing wards 20 and 21. In the bin location plan, 42 bins are located in the market area. Sizes of the bins were calculated by dividing the total waste generated in Gole Bazar by the total number of bins in the market area (with FOS). The composition for source segregation is taken as 50% recyclables and 50% mixed waste.

Total commercial waste in Kharagpur = 52.4 MT
Gole Bazar is the biggest market in Kharagpur and contributes 50% of the total commercial waste in the city
Waste generated in Gole Bazar = 26.2 MT/day

(continued)

There are 42 bin locations in Gole Bazar,
Therefore, the weight of waste in each bin location = $26.22 \times 10^3/42 = 624.29$ kg/day
Volume of waste in each bin location = (624.29 kg/day)/(250 kg/m^3) = 2.5 m^3
Size of mixed waste bin = $2.5 \times 0.5 \times 2$ (FOS) = 2.5 m^3
Size of recyclables bin = $2.5 \times 0.5 \times 2$ (FOS) = 2.5 m^3

4.2 Site for Transfer Station (TS) and Landfill

As per recent SWM Rules 2016, if disposal sites are more than 15 km away from the collection area, it is economical to set up transfer stations to save transportation time and fuel (MSWM Manual, CPHEEO 2016). The maximum distance within the collection area to the landfill is from Ward 12 which is around 23 km and a transfer station is likely to be cost-effective.

A site was identified for a transfer station using Google Earth Pro and confirmed by a site visit. The site selected is located near the centre of the city with an area of 1.0 hectare (Fig. 8.6). This will help in decreasing collection costs by reducing travelling time and distance from the service area to the landfill. Also there is extra space available to construct Material Recovery Facility (MRF) to segregate recyclable or there is enough space available to build a treatment facility like a bio-gas or a compost plant (Outlined in yellow colour in Fig. 8.6).

The transfer station site is adjacent to a railway line allowing transfer of recyclables to other cities for treatment. The selected site is about 1 km away from State Highway (SH-5) and 8.6 km away from Kolkata-Mumbai highway (NH-49). Due to proximity from SH-5, heavy vehicles can be used for transportation of waste from

Fig. 8.6 Site for transfer station

Fig. 8.7 Route from transfer station to landfill (17.2 km route length)

transfer station to landfills resulting in reduced transportation costs by reducing the number of trips (Fig. 8.7).

4.3 *Landfill Size Calculation and Site Selection*

The next step was to find a suitable location for the landfill. River Kansabati lies to the north of the city and on the west side of the city is the Kalaikunda Airbase of the Indian Air Force. Further, there are villages and vegetation on unused land to the south of the city. So, the only option left for a landfill site is to the west of the city as shown in Fig. 8.8.

The current site was selected according to CPHEEO guidelines of the Indian government and is at a considerable distance from the city to provide for future expansion (Fig. 8.8). Assumptions made for selecting the landfill site and calculating its size are:

- The landfill site is designed for filling of waste for 20 years.
- Only 15% of the total waste collected will be landfilled.
- The per capita waste generation rate is assumed to be constant over a period of a year. The literature shows that as economic conditions of the region or country improve, living standards of the people also increase resulting in higher per capita waste generation rates (Goel 2008). Therefore, it is assumed that per capita waste generation rate increases by 1.5 times every decade. So, in 20 years it will be three times the current per capita waste generation, i.e., in 2017, and is likely to increase to a maximum of 1.5 kg/capita-day.
- To calculate waste generated in each year, population data are needed. So, exponential growth rate method was used for population forecasting.

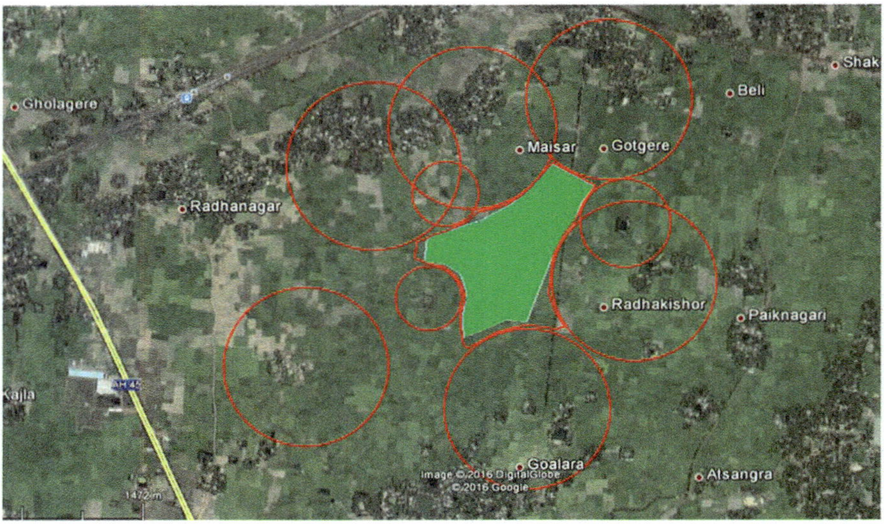

Fig. 8.8 Site for landfill based on regulations (larger circles are of radius 500 m and smaller circles are of radius 300 m showing site is minimum 500 m and 300 m away from residential areas and ponds, respectively)

- Due to shallow groundwater depth, height of the landfill is assumed to be 10 m of which 5 m is below and 5 m above the ground level.
- The landfill area used for design is 1.2 times the required area. In the extra 20% area, 10% extra is provided for buildings, roads, monitoring labs, leachate treatment plant and 10% extra area is provided as per CPHEEO guidelines.

The total waste generated in 20 years is estimated to be approximately 2.7 million metric tons of which 0.4 million metric tons of waste is to be landfilled (15% of total waste). Area required for the landfill is estimated to be 6.93 hectares. Landfill size calculation is shown below.

Total waste coming to landfill in 20 years = 404,358.42 MT (Table 8.7)
Compaction density = 700 kg/m^3
Volume of waste coming to the landfill = 577,654.88 m^3
Height of the landfill = 10 metres (2 m extra for landfill cover)
Area required = 577,654.88/10 = 57,765.488 m^2
20% extra area provided = 10% extra (CPHEEO) + 10% extra (building, treatment, others)
Net area required for landfill = 1.2 × 57,765.49 = 69,320 m^2 (approx.)

4.4 *Vehicle Routing*

In developing countries like India, waste collection and transportation comprise approximately 80–90% of the overall SWM budgets while collection efficiencies remain less than half. In contrast, collection efficiencies are greater than 90% in high-income countries and collection costs are less than 10% of the total SWM budgets (Hoornweg and Bhada-Tata 2012). Therefore, optimization of the collection routes and bin locations is a critical need in developing countries.

Vehicle routing can be done manually or with tools such as remote sensing, GIS and linear programming for any area. Best possible vehicle routes were drawn manually for the market area and compared with routes obtained using linear programing. Further, these routes can be monitored using global positioning systems (GPS) for automated control and data collection (Wilson et al. 2007).

4.4.1 Assumptions and Basic Requirements for Vehicle Routing

- If the system is *semi-automated collection*, then it typically requires a crew of two: a driver and a collector.
- Route starts from garage (at municipality office) and after collection goes to transfer station. All following routes on the same day begin from the transfer station and return to it. At the end of the work day, the collection vehicle returns to the garage from transfer station.
- Average velocity of truck on road is 25 km/h while it is 15 km/h during collection.
- Trucks provided are of sizes 19 m^3 and 11 m^3 (available in the market).
- Truck compartment ratio of 70:30 was assumed for residential areas assuming mixed waste as 70% of total volume and recyclables as 30% of total volume.
- Trucks used for collecting commercial waste were assumed to have two compartments: one for mixed waste (50% of volume) and second is for recyclables (50% of volume).
- Density of compacted waste in the truck is 590 kg/m^3.
- Collection time at a location: 1.5 min.
- Unloading time for truck at transfer station was 10 min (with factor of safety; otherwise 7.5 min maximum).
- 25% extra time was provided for off-route time to account for traffic problems, bad weather conditions, vehicle problems and personal requirements.

4.4.2 Calculation of Locations Served Per Trip

To design a collection system with 100% efficiency, the number of bins in each ward, the volume of waste collected in each bin and the truck capacity are required. So, number of locations/bins served are calculated below for each ward, market area and apartment buildings.

Truck size – 19 m^3

Volume available for mixed waste = 70% of total volume = 0.7 × 19 m^3 = **13.3 m^3** = 13.3 m^3 × 590 kgm^{-3} = **7847 kg**

Volume available for recyclables = 30% of total volume = 0.3 × 19 = **5.7 m^3** = 5.7 m^3 × 590 kgm^{-3} = **3363 kg**

For Ward 20

Mixed waste: Bin size = 1.25 m^3 (Volume); 1.25 m^3 × 250 kg m^{-3} = 312.5 kg (weight)

Locations served/trip = Maximum weight of waste/weight of bin = 7847 kg/312.5 kg = 25 locations

Recyclables: Bin size = 0.5 m^3 (Volume); 0.5 m^3 × 250 kg m^{-3} = 125 kg (weight)

Locations served/trip = Maximum weight of waste/weight of bin = 3363 kg/125 kg = 26 locations

Maximum locations served in a trip are ***25 locations per trip***

Similarly, maximum locations served per trip are calculated for Ward 21, market area and apartment area

For Ward 21: *Maximum locations served per trip are* ***19 locations per trip***

For market: *Maximum locations served per trip are* ***nine locations per trip***

For 6-floor apartment buildings: *Maximum locations served per trip are* ***31 locations per trip***

For 3-floor apartment buildings: *Maximum locations served per trip are* ***52 locations per trip***

4.4.3 Manual Vehicle Routing

In general, vehicles are routed manually. However for large areas, manual methods can result in inefficient utilization of resources and higher collection costs (when compared to other methods). Using tools/software like Google Earth and specific guidelines, routing efficiency can be increased.

The following guidelines were used for developing collection routes manually (Worrell and Vesilind 2012):

- Routes should not overlap, should be compact, and should not be fragmented.
- The starting point should be as close to the truck garage or transfer station as possible.
- One-way streets that cannot be traversed in one line should be looped from the upper end of the street.
- Dead-end streets should be collected from the left side of the street since most vehicles in India are 'right hand drive'.
- Anti-clockwise turns around blocks should be used whenever possible.
- Long, straight paths should be routed before looping anti-clockwise.

- Routes should be laid out so that the last container to be collected on the route is located nearest to the disposal site.
- Existing policies and regulations related to such items as the point of collection and frequency of collection must be identified.
- Existing systems, such as crew size and vehicle types, must be coordinated.
- Wastes generated at traffic-congested locations should be collected as early in the day as possible or even at night.

4.4.4 Routes for Residential and Apartment Area

Routes were developed using Google Earth software and time was calculated from route length and average velocity of the truck. Routes are drawn to collect waste generated in the residential area, apartment area and special bins like the police office, and schools. All routes start from the garage in the morning which is located in the municipality office complex and end at the garage. The first route starts from the garage and then it goes to the entry point in the service ward. After that, waste is collected from the service ward/area till the truck is completely filled. After it is completely filled, the truck goes to the transfer station to unload its contents. The next route will start from the transfer station (TS), return to the next collection point in the ward and start collecting from bins. After it is completely filled, it will again come back to the TS for unloading. At the end of the day, the truck will return to the garage from the TS.

A total of six routes were planned for waste collection in the residential and apartment area. The six routes have a total distance of 63.91 km and the time required for these routes is 603 min (including unloading and 25% extra time). Since the time taken for completing the routes is more than 8 h, two collection trucks will be required (routes distance and time are shown in Table 8.1). Calculation for the collection time required for NR_1 is shown below and the route is shown in Fig. 8.9.

Route 1: NR_1 (10 stops and 34 bins locations)
Garage to entry point: 4.8 ≈ 5 min [Distance: 2 km (with 25 km/h speed)]
Entry point to first bin: 2.4 ≈ 2.5 min [Distance: 1 km (with 25 km/h speed)]
Picking up time: 51 min [34 locations × 1.5 min]
Travelling time: 8.5 min [Route length: 2.13 km (with 15 km/h speed)]
From starting point to TS: 6.24 ≈ 6.5 min [Distance: 2.6 km (with speed 25 km/h)]
Time taken: 73.5 min + 18.38 min (25% extra) ≈ 92 min
Distance covered: 7.73 km (NR_1 shown in Fig. 8.9)

Using the same methodology, time was calculated for the remaining five routes and the results are summarized in Table 8.1.

Table 8.1 Time and distance of six routes covering household areas in wards

Routes	Routing time (including 25% extra time), min	Unloading time, min	Total time, min	Distance, km
NR_1	92	10	102	7.73
NR_2	102	10	112	9.04
NR_3	87	10	97	9.72
NR_4	82	10	92	10.77
NR_5	86	10	96	12.02
NR_6	94	10	104	14.63
Total	543	60	603	63.91

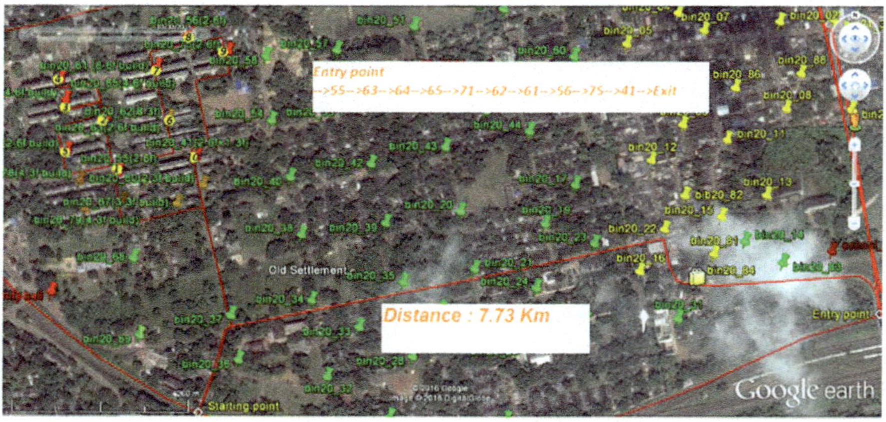

Fig. 8.9 Route NR_1

4.4.5 Routes for Commercial/Market Area

The market area is a very busy place so the quantity of waste produced is large and varies from day-to-day. Two possible options for collecting market waste are: collect all commercial bin waste once or twice a day. The method that is most economical in terms of efficiency and cost is the best choice. Methodology for planning vehicle routes for both options is the same as the methodology used for residential and apartment areas.

For a collection frequency of once a day, a total of six routes were planned for waste collection in the market area (Fig. 8.10). The six routes have a total distance of 54.79 km (shown in Table 8.2) and time required for these routes is 365 min (including unloading and 25% extra time). The time taken for completing the routes is less than 8 h, so only one truck per day is required for collection.

For collection frequency of twice a day, a total of three routes were planned in the market area (Fig. 8.11). These three routes have a total distance of 60.96 km (after two-time collection) and time required for these routes is 468 min (shown in

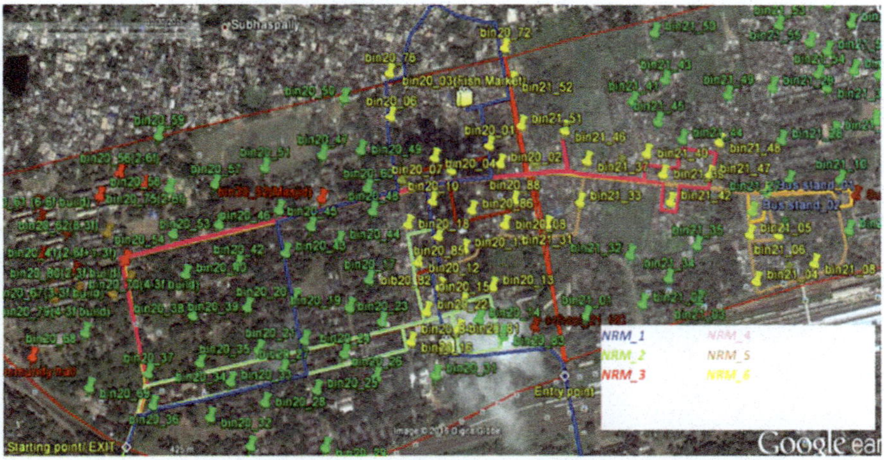

Fig. 8.10 Collection routes for market area (one time collection)

Table 8.2 Time and distance of six routes covering commercial area in wards (one time collection)

Routes	Routing time (including 25% extra time), min	Unloading time, min	Total time, min	Distance, km
RM_1	46	10	56	7.45
RM_2	46	10	56	7.73
RM_3	50	10	60	8.38
RM_4	53	10	63	9.15
RM_5	56	10	66	9.71
RM_6	54	10	64	12.37
Total	305	60	365	54.79

Table 8.3). The time taken for completing the routes is less than 8 h, so only one collection truck is required per day.

On comparing both cases, time taken and distance covered in collection is more with collection frequency of two times a day compared to one time a day. This implies that labour and fuel costs will be greater with two times a day collection rather than one time a day collection. Therefore, one time a day collection is more cost-effective and will provide 100% collection efficiency (Table 8.4).

4.4.6 Optimization of Routes in the Market Area Using Linear Programming (LP)

Another approach for route optimization is to use the same algorithm as the *travelling salesman problem (TSP)*. Travelling salesman problem is defined as "If a list of nodes and distances between each pair of nodes is given, what is the shortest

Fig. 8.11 Collection routes for market area (two times collection)

Table 8.3 Time and distance of six routes covering commercial area in wards (two times collection)

Routes	Routing time (in min) (Including 25% extra time)	Unloading time (in min)	Total time (in min)	Distance (in km)
RM2_1	61	10	71	7.5
RM2_2	72	10	82	9.82
RM2_3	71	10	81	13.16
Total	204	30	234	30.48

Table 8.4 Cost analysis of one and two times a day collection of waste in market area (for 20 years)

Factors	Case a	Case b
Bin Volume	2.5 m^3	1.25 m^3
Bin price	Rs 20,000	Rs 13,000
Price of all bins	42 × 20,000 = 8.4 lakhs	42 × 13,000 = 5.46 lakhs
Fuel (CNG) required in a year (truck mileage 4 km/L)	365 × 54.79 km/4 km/L = 4999.6 L	365 × 60.96 km/4 km/L = 5562.6 L
Fuel price (On 01-11-2017)	Rs 43.6 per L	Rs 43.6 per L
Cost due to fuel in a year	Rs. 4999.6 × 43.6 = 2.18 lakhs/year	Rs. 5562 × 43.6 = 2.43 lakhs/year
Cost for 5 years	(5 × 2.18) + 8.4 = 19.3 lakhs	(5 × 2.43) + 5.46 = 17.61 lakhs
Cost for 10 years	30.2 lakhs	29.76 lakhs
Cost for 20 years (Design life)	52 lakhs	54.06 lakhs

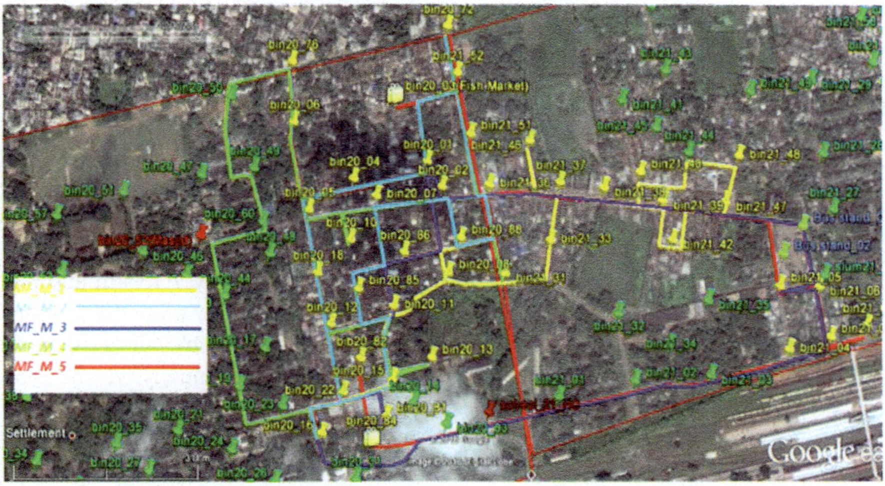

Fig. 8.12 Optimized collection routes based on linear programming for the market area

Table 8.5 Time and distance covered in five optimized routes in the commercial area

Routes	Routing time (including 25% extra time) (in min)	Unloading time (in min)	Total time (in min)	Distance (in km)
MF_M_1	47	10	57	8.17
MF_M_2	47	10	57	8.52
MF_M_3	50	10	60	9.11
MF_M_4	48	10	58	9.17
MF_M_5	60	10	70	12.69
Total	252	60	**302**	**47.66**

possible route that will visit each and every node exactly one time and at the end return to the origin". Assuming each bin is a node, TSP ensures collection from all bins while minimizing distance covered. Two constraints were applied for optimization, no value can be negative, and a maximum of nine nodes can be covered in a single route. The second constraint was applied because in the market area, a maximum of nine bins can be picked up in a single route due to truck capacity limitation. In TSP, the route begins and ends at bin 20_16 in Ward 20. Bin 20_16 is selected because it is closest to both garage and TS while entering in the market area for collection. Distance of bin 20_16 from TS and garage was calculated using Google Earth and included in route planning. The linear program code was written in Visual C++ and used to determine optimum routes.

After optimization, there are five collection routes for the market area (shown in Fig. 8.12). These five routes have a total distance of 47.66 km (shown in Table 8.5) and time required for these routes is 302 min (including unloading and 25% extra

Table 8.6 Comparison between manually developed collection routes and those based on linear programming (LP)

Parameter	Manually	After optimization	Reduction after optimization
Time	6 h 5 min	5 h 2 min	63 min
Distance	54.79 km	47.66 km	7.13 km

time). The time taken for completing the routes is less than 8 h, so for collection only one truck per day is required.

Optimized routes result in reducing total routing distance by 7.13 km and total vehicle routing time by approximately 1 h (63 min). After optimization, there are five routes instead of six resulting in significant savings in fuel, time and cost in the collection process. A comparison of the results obtained manually versus those obtained using LP is provided in Table 8.6.

5 Conclusions

An integrated solid waste management (ISWM) plan was developed for Kharagpur using RS, GIS and linear programming. Major conclusions of this study are:

1. A bin location and sizing plan was developed using Google Earth Pro and guidelines provi ded by the Government.
2. Sites for a transfer station and landfill were identified using Google Earth Pro and their areas calculated

Using Google Earth Pro, it was possible to route collection vehicles manually based on ground/local information or using a linear programming approach.

These methods were applied successfully at very low materials and labour costs since satellite images and Google Earth Pro are freely available from the Internet. These methods can be extended further for detailed project monitoring, cost estimation and cost-benefit analyses of alternatives for developing an ISWM plan for any other city or region.

Acknowledgements The authors are grateful to officials and staff of Kharagpur Municipal Corporation for providing data. The authors are also very grateful to Mr. Anurag Anand for writing the linear programming code for optimization of the vehicle collection routes.

Appendix

Table 8.7 Calculation of the total amount of waste going for landfilling from 2016 to 2036

Year	Population	SW gen. rate (kg/capita-day)	Total waste generation/day (in kg)	Total waste going to landfill (in kg/day) (15% of total waste)	Landfill waste (kg/year)	Volume of waste for landfill (p = 700 kg/m^3)
2016	314,628	0.5	157,314	23,597.1	8,636,538.6	12,337.91229
2017	317,015	0.55	174,358.25	26,153.7375	9,546,114.188	13,637.30598
2018	320,000	0.6	192,000	28,800	10,512,000	15,017.14286
2019	323,013	0.65	209,958.45	31,493.7675	11,495,225.14	16,421.7502
2020	326,055	0.7	228,238.5	34,235.775	12,530,293.65	17,900.4195
2021	329,125	0.75	246,843.75	37,026.5625	13,514,695.31	19,306.70759
2022	332,224	0.8	265,779.2	39,866.88	14,551,411.2	20,787.73029
2023	335,352	0.85	285,049.2	42,757.38	15,606,443.7	22,294.91957
2024	338,510	0.9	304,659	45,698.85	16,725,779.1	23,893.97014
2025	341,697	0.95	324,612.15	48,691.8225	17,772,515.21	25,389.30745
2026	344,915	1	344,915	51,737.25	18,884,096.25	26,977.28036
2027	348,162	1.05	365,570.1	54,835.515	20,014,962.98	28,592.80425
2028	351,441	1.1	386,585.1	57,987.765	21,223,521.99	30,319.31713
2029	354,750	1.15	407,962.5	61,194.375	22,335,946.88	31,908.49554
2030	358,090	1.2	429,708	64,456.2	23,526,513	33,609.30429
2031	361,462	1.25	451,827.5	67,774.125	24,737,555.63	35,339.36518
2032	364,866	1.3	474,325.8	71,148.87	26,040,486.42	37,200.69489

(continued)

Table 8.7 (continued)

Year	Population	SW gen. rate (kg/capita-day)	Total waste generation/day (in kg)	Total waste going to landfill (in kg/day) (15% of total waste)	Landfill waste (kg/year)	Volume of waste for landfill (p = 700 kg/m^3)
2033	368,301	1.35	497,206.35	74,580.9525	27,222,047.66	38,888.63952
2034	371,769	1.4	520,476.6	78,071.49	28,496,093.85	40,708.7055
2035	375,270	1.45	544,141.5	81,621.225	29,791,747.13	42,559.63875
2036	378,803	1.5	568,204.5	85,230.675	31,194,427.05	44,563.46721
				Total waste	404,358,414.90	577,654.89

References

Annepu, R.K. (2012). Sustainable Solid Waste Management in India. Department of Earth and Environmental Engineering at Columbia University.

CPHEEO (2016). Report on "Manual on Municipal Solid Waste Management" http://cpheeo.nic.in/SolidWasteManagement2016.htm.

Dutta, D. and Goel, S. (2017). Applications of Remote Sensing and GIS in Solid Waste Management – A Review. *In:* Advances in Solid and Hazardous Waste Management, 129–150. Springer & Capital Publishing Co.

Goel, S. (2008). Municipal Solid Waste Management (MSWM) in India: A Critical Review. *Journal of Environmental Science and Engineering, NEERI*, **50(4):** 319–328.

Gupta, S., Mohan, K., Prasad, R., Gupta, S. and Kansal, A. (1998). Solid Waste Management in India: Options and Opportunities. *Resources, Conservation and Recycling*, **24(2):** 137–154. Doi: https://doi.org/10.1016/S0921-3449(98)00033-0.

Hoornweg, D. and Bhada-Tata, P. (2012). What a Waste: A Global Review of Solid Waste Management. Urban Development Series Knowledge Papers. Washington DC, USA. doi: https://doi.org/10.1111/febs.13058.

Naresh, K. and Goel, S. (2009). Characterization of Municipal Solid Waste (MSW) and a Proposed Management Plan for Kharagpur, West Bengal, India. *Resources, Conservation and Recycling*, **53(3):** 166–174. doi:https://doi.org/10.1016/j.resconrec.2008.11.004.

Swachh Bharat Mission (2016). Municipal Solid Waste Management Manual – Part I: An Overview.

Wilson, B.G., Betsy, J.A., Brian, W.B. and Winning, A. (2007). Practical Applications for Global Positioning System Data from Solid Waste Collection Vehicles. *Canadian Jour. Civil Eng.*, **681:** 678–681. doi:https://doi.org/10.1139/L06-174.

Worrell, W.A. and Vesilind, P.A. (2012). Solid Waste Engineering. 2nd Edition. Stamford, CT, USA: Cengage Learning.

Chapter 9
Optimization of F/M Ratio During Anaerobic Codigestion of Yard Waste with Food Waste: Biogas Production and System Stability

Sagarika Panigrahi, Hari Bhakta Sharma, and Brajesh K. Dubey

1 Introduction

As the two major organic fractions of municipal solid waste, food waste (FW) has lower carbon to nitrogen (C/N) ratio and yard waste (YW) has high C/N ratio. Carbon acts as the energy source for the microorganisms, whereas nitrogen plays a major role in increasing the microbial population. The C/N ratio indicates the nutrient level of a feedstock. Microorganisms need a suitable ratio of C/N for their metabolic process, and thus, anaerobic digestion (AD) process is sensitive to C/N ratio. A low C/N ratio may cause excess ammonia inside the digester, which may hamper the degradation of organic matter and may inhibit methanogens (Brown and Li 2013). In this scenario, co-digestion of FW and YW is a new opportunity for the C/N balance during AD. In most of the literature it has been reported that C/N ratio should be in the range of 20:1–30:1 with a fixed ratio of 25:1 optimum for bacterial growth in an AD system (Khalid et al. 2011).

Beside the C/N ratio another crucial factor affecting AD process is the food to mass (F/M) ratio expressed as the amount of feedstock added per amount of inoculum on volatile solid (VS) basis (Cheng and Zhong 2014). Theoretically F/M ratio can affect the kinetics of AD process, and it doesn't have any effect on ultimate methane yield (which depends only on the food or feed stock). However, experimental results have concluded that high F/M ratio can be toxic and low F/M ratio can inhibit enzyme induction for organic matter degradation or fasten the startup period to obtain higher experimental methane yield (Cheng and Zhong 2014; Krishania et al. 2013). In this situation, it is important to minimize the requirement of active inoculum and avoid the toxic condition inside the digester. Previous studies on F/M ratio optimization during AD of organic fraction of municipal solid waste were

S. Panigrahi · H. B. Sharma · B. K. Dubey (✉)
Civil Engineering Department, Indian Institute of Technology Kharagpur, Kharagpur, West Bengal, India
e-mail: bkdubey@civil.iitkgp.ac.in

D. Sengupta et al. (eds.), *Treatment and Disposal of Solid and Hazardous Wastes*,
https://doi.org/10.1007/978-3-030-29643-8_9

complementary. In the study of Neves et al. (2004) F/M ratio of 0.5–2.3 was recommended for AD process whereas in the study of Brown and Li (2013), a decrease in biogas production was observed with increase in F/M ratio from 1.6 to 5.0. In the study of Elbeshbishy et al. (2012), optimization of F/M ratio for AD of FW was conducted at a constant pH 7. However, maintaining a constant pH externally may not be feasible. Based on the above research gaps, this work investigates about the optimization of F/M ratio during anaerobic co-digestion of FW and YW under an uncontrolled pH.

In the present research, a lab scale batch anaerobic co-digestion of FW and YW was conducted. The purpose of co-digestion was (i) optimize F/M ratio during co-digestion based on biogas production; and (ii) effect of different F/M ratio on reactor stability during anaerobic co-digestion.

2 Material and Methodology

2.1 Collection and Characterization of Yard Waste, Food Waste and Inoculum

The YW was collected from the Indian Institute of Technology Kharagpur campus, India, Inorganic contaminants (debris, plastic, metal contaminants etc.) were separated manually. The quartering method was adopted to collect the representative sample and the representative sample mainly composed of grass (33% by wt.), dry leaves (33%) and fallen sticks (2%). The collected YW was grounded by a domestic mixture and stored in an air-tight plastic bag for further use. FW used in this study was collected from girls' hostel canteen. The collected FW mainly composed of cooked rice, vegetables, bread, and meat. Then the collected FW was homogenised by a domestic mixture grinder and stored at a temperature of 4 °C, for further use. Fresh waste activated sludge (WAS) was collected from the mesophilic AD plant, running inside the campus. The collected WAS was centrifuged to increase the total solid (TS) content from 6.7% to 10.5%. The collected WAS was used as an inoculum for batch AD study. The initial characteristics of YW, FW and WAS is presented in Table 9.1.

Table 9.1 Composition of substrate and inoculum used for batch anaerobic digestion

F/I ratio	Yard waste (g)	Food waste (g)	Total amount of feedstock (g)	Inoculum (g)	Total weight (g)
1.0	36	108	144	144	288
1.5	43	130	173	115	288
2.0	48	144	192	96	288
2.5	51	154	206	82	288

2.2 Batch Biochemical Methane Potential Study

All substrates and inoculum (YW, FW and WAS) were added into the digestion reactor, the proportion of feedstock is mentioned in Table 9.1. The composition of FW and YW were selected in order to maintain a C/N ratio of 25.0, calculated using Eq. 9.1. The proportion of feedstock (YW and FW) and WAS was calculated based on Eq. 9.2. Then nitrogen gas was purged into the reactor for 4 min to create a complete anaerobic condition. Soon after that the digestion reactors were connected to the displacement tank filled with 1.5 N NaOH through a plastic pipe. A collector tank was arranged with the displacement tank (Fig. 9.1) to collect the displaced water from displacement tank (Esposito 2012). Thymol blue was added to the NaOH solution and used as an indicator of alkalinity of the solution. The batch AD experiment was conducted for 30 days until the cumulative biogas production reached a stable value.

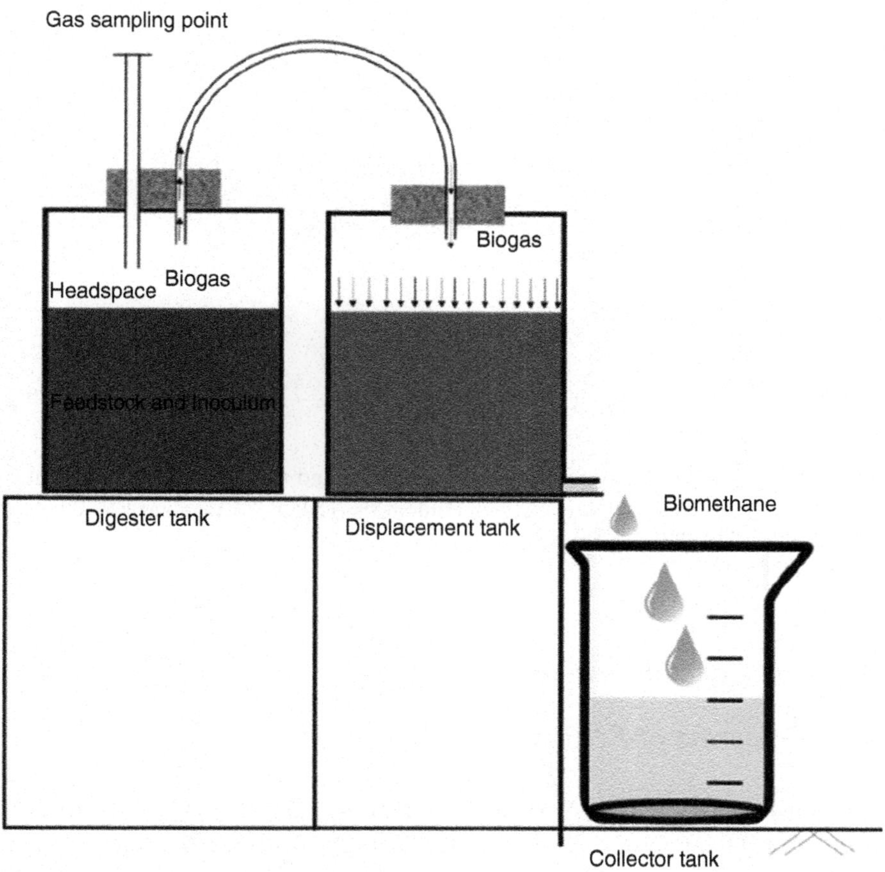

Fig. 9.1 Batch biochemical methane potential setup for codigestion of yard waste and food waste

2.3 Analytical Methods

The TS, VS, pH and soluble COD (sCOD) were measured according to the standard protocols of the American Public Health Association (APHA 1995). Volatile fatty acid (VFA) was measured by the direct titration method with 0.1 N H_2SO_4 up to pH 3.3 and then 0.1 N NaOH up to pH 7.0, according to the procedure described by DiLallo and Albertson (1961). The methane production was measured daily by adopting water displacement method and substrate characteristics in the reactor were measured after every 6 days during 30 days of digestion period. Compositional analysis: cellulose (Updegraff 1969), hemicellulose (Goering and Van 1975), lignin (National renewable energy laboratory procedure) (Sluiter et al. 2008) contents of the untreated and pretreated reactors were measured. Ultimate analysis in terms of carbon (C), hydrogen (H) and nitrogen (N) contents and ratio of C/N was carried out using standard procedure of CHNS analysis by using a fully automatic elemental analyser (EURO EA, Germany).

2.4 Calculation Methods

The total C/N of the mixed material was calculated using Eq. (9.1):

$$(\mathrm{C/N})_{\mathrm{Mixture}} = \frac{W_{\mathrm{FW}} \times C_{\mathrm{FW}} + W_{\mathrm{YW}} \times C_{\mathrm{YW}}}{W_{\mathrm{FW}} \times N_{\mathrm{FW}} + W_{\mathrm{YW}} \times N_{\mathrm{YW}}} \tag{9.1}$$

F/M was calculated with Eq. (9.2):

$$\frac{F}{M} = \frac{W_{\mathrm{FW}} \times VS_{\mathrm{FW}} + TS_{\mathrm{YW}} \times VS_{\mathrm{YW}}}{VS_{\mathrm{WAS}}} \tag{9.2}$$

where W_{FW} and W_{YW} are the amount (g TS); C_{FW} and C_{YW} are the carbon content (% TS); N_{FW} and N_{YW} are the nitrogen content (% TS); and VS_{FW} and VS_{YW} are the VS content (% of TS), in FW and YW respectively.

3 Result and Discussion

3.1 Properties of Feedstocks

The initial characteristics of YW, FW and WAS used in this study for experimental methane yield are given in Table 9.2. The high initial TS and VS content in YW (89.5 ± 0.6%, 97.3 ± 2.2% TS), and FW (22.6 ± 0.8, 99.8 ± 0.1% TS), made the substrates suitable for solid state AD. Although the initial pH of FW is below 4.0,

Table 9.2 Initial characteristics of yard waste (YW), food waste (FW) and waste activated sludge (WAS)

Parameter	Unit	YW	FW	WAS
TS	%	89.5 ± 0.6	22.6 ± 0.8	10.0 ± 2.5
VS	% TS	97.3 ± 0.6	99.8 ± 0.1	99.2 ± 0.1
pH		7.5 ± 0.05	3.7 ± 0.7	7.8 ± 0.46
VFA	mg/L	363.75 ± 20.16	970 ± 127	157.45 ± 13.16
sCOD	mg/g TS	189.35 ± 15.45	634.8 ± 19	216.30 ± 24.92
C	%	46.2	42.7	
H	%	5.83	9.1	
O	%	46.9	46.2	
N	%	1.03	1.98	
S	%	< 0.3	0.3	
C/N		44.8	21.6	9.43

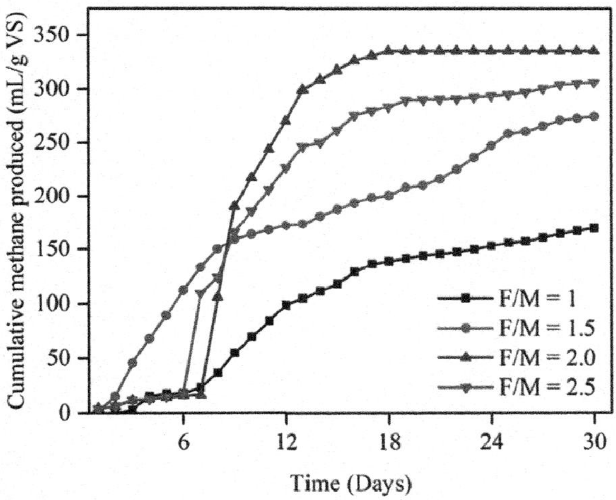

Fig. 9.2 Cumulative methane yield for different F/M ratios during anaerobic co-digestion of food waste and yard waste

initial pH of the YW and WAS were 7.5 ± 0.05 and 7.8 ± 0.46%. The initial C/N ratio of YW and FW were 44.83 and 21.57, respectively. In order to maintain an optimum C/N ratio (25.0) YW and FW were mixed in a proportion of 1:3.

3.2 Effect of F/M Ratio on Methane Yield

The effect of four different F/M ratios (1.0, 1.5, 2.0 and 2.5) on cumulative methane production during anaerobic co-digestion of FW and YW is shown in Fig. 9.2. The biogas production in the initial days was low due to hydrolysis of complex organic matter. The higher initial biogas production for low F/M ratios (F/M ratios 1.0 and 1.5) was likely due to the high amount of WAS, which provided relatively large

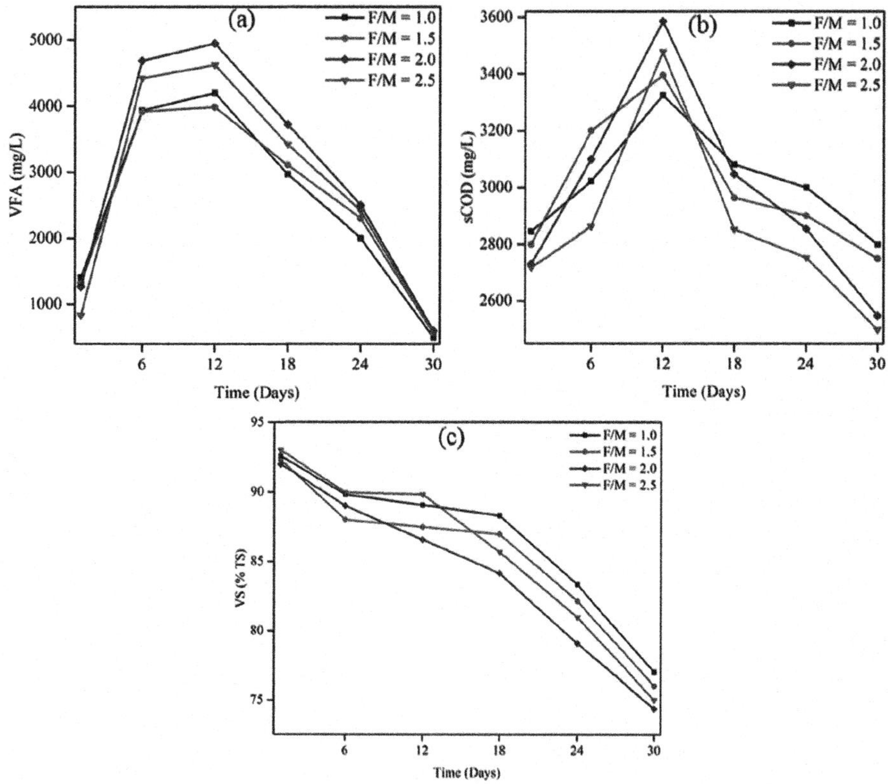

Fig. 9.3 Variation in VFA (**a**), sCOD (**b**), and VS (**c**) for different F/M ratios during anaerobic co-digestion of food waste and yard waste

concentrations of hydrolytic enzymes, fermenting and methanogenic bacteria. After initial lag phase, the biogas production was increased for F/M ratios 2.0 and 2.5, and the production rate was higher than F/M ratios 1.0 and 1.5. The cumulative experimental methane yield was: 169 mL/g VS (F/M = 1.0), 274 mL/g VS (F/M =1.5), 335 mL/g VS (F/M = 2.0), and 306 mL/g VS (F/M = 2.5).

3.3 *Effect of F/M Ratio on VFA Production*

The variation of VFA obtained during anaerobic co-digestion of YW and FW for different F/M ratios is presented in Fig. 9.3a. The initial VFA concentration was 1412.75 mg/L for F/M ratio of 1.0, 1325.41 mg/L for F/M ratio of 1.5, 1268.75 mg/L for F/M ratio of 2.0, and 831.25 mg/L for F/M ratio of 2.5. The initial VFA concentration was found decreasing with increasing F/M ratio, due to increase in inoculum concentration. A rapid increase in VFA production was noticed within 6 days; this is due to conversion of organic matter into simpler monomeric compounds (Elbeshbishy et al. 2012). The VFA production was increased up to 12th day, and the highest VFA production (4950.45 mg/L) was found for F/M ratio 2.0. The decrease in VFA concentration after 12th day can be marked as the initiation of methanogenesis stage. The VFA concentration throughout the digestion period was lower than the inhibiting concentration i.e., 13,000 mg/L indicating favourable conditions for the growth and activity of acid-producing microorganisms.

3.4 Effect of F/M Ratio on Organic Matter Solubilisation

The variation in sCOD throughout digestion period is depicted in Fig. 9.3b. The initial sCOD was high (2844.44 mg/L) for low F/M ratio (1.0) due to higher fraction of inoculum and further increase in F/M ratio led to decrease in sCOD to 2804.15 mg/L for F/M ratio 1.5, 2730.08 mg/L for F/M ratio 2.0 and 2718.3 mg/L for F/M ratio 2.5. The sCOD concentration was increased on 6th day, achieved a highest value on 12th day and after that sCOD started decreasing. The highest sCOD was found for F/M ratio 2.0 (3584.52 mg/L) followed by 2.5 (3478.87 mg/L), 1.5 (3395.74 mg/L), and 1.0 (3325.45 mg/L). The final sCOD measured on 30th day was higher for F/M ratios 1.0 and 1.5 in comparison to F/M ratios of 2.0 and 2.5, which indicates digester instability for F/M ratios 1.0 and 1.5.

3.5 Effect of F/M Ratio on VS Degradation

The degradation in VS throughout the study period is presented in Fig. 9.3c. As expected, no significant difference in initial VS was observed for different F/M ratios because the F/M ratio was maintained based on the VS content. VS is an important parameter in AD; the amount of VS removed is directly related to biogas production (Zhu et al. 2010). A sharp decrease in VS content in the initial 6 days is related to removal of easily biodegradable structural components. The rate of decrease was quite moderate between 6th and 12th days, may be due to hydrolysis of structural components (cellulose, hemicellulose and lignin). After 18th day, a sharp decrease in VS reduction was noticed, which is related to the methanogenesis phase. A highest VS reduction (74.38% TS) was noticed for F/M ratio of 2.0 indicating highest amount of biogas production, which was also confirmed from Fig. 9.2.

4 Conclusion

The present study made an effort to handle two major organic fractions of municipal solid waste (food waste and yard waste) through anaerobic co-digestion. Optimization of F/M ratio was also done to improve biogas production. Highest biogas production was observed for F/M ratio of 2.0. As expected, the highest organic matter solubilisation, VFA production and VS degradation were also observed for the same F/M ratio.

References

APHA (1995). WPCF, Standard methods for the examination of water and wastewater. Am. Public Heal. Assoc. Washington, DC.

Brown, D. and Li, Y. (2013). Solid state anaerobic co-digestion of yard waste and food waste for biogas production. *Bioresour. Technol.*, **127:** 275–280.

Cheng, X.Y. and Zhong, C. (2014). Effects of feed to inoculum ratio, co-digestion, and pretreatment on biogas production from anaerobic digestion of cotton stalk. *Energy and Fuels*, **28:** 3157–3166.

DiLallo, R. and Albertson, O.E. (1961). Volatile acids by direct titration. *J. Water Pollut. Control Fed.*, 356–365.

Elbeshbishy, E., Nakhla, G. and Hafez, H. (2012). Biochemical methane potential (BMP) of food waste and primary sludge: Influence of inoculum pre-incubation and inoculum source. *Bioresour. Technol.*, **110:** 18–25.

Esposito, G. (2012). Bio-Methane Potential Tests to Measure the Biogas Production from the Digestion and Co-Digestion of Complex Organic Substrates. *Open Environ. Eng. J.*, **5:** 1–8.

Goering, H. and Van, S. (1975). Forage fibre analysis (Ed.). UDoAR Service.

Khalid, A., Arshad, M., Anjum, M., Mahmood, T. and Dawson, L. (2011). The anaerobic digestion of solid organic waste. *Waste Manag.*, **31:** 1737–1744.

Krishania, M., Vijay, V.K. and Chandra, R. (2013). Methane fermentation and kinetics of wheat straw pretreated substrates co-digested with cattle manure in batch assay. *Energy*, **57:** 359–367.

Neves, L., Oliveira, R. and Alves, M.M. (2004). Influence of inoculum activity on the bio-methanization of a kitchen waste under different waste/inoculum ratios. *Process Biochem.*, **39:** 2019–2024.

Sluiter, A., Ruiz, R., Scarlata, C., Sluiter, J. and Templeton, D. (2008). Determination of Extractives in Biomass: Laboratory Analytical Procedure (LAP). Issue Date 7/17/2005 - 42619.pdf. Tech. Rep. NREL/TP-510-42619 1–9. https://doi.org/NREL/TP-510-42621

Updegraff, D.M. (1969). Semimicro determination of cellulose in biological materials. *Anal. Biochem.*, **32:** 420–424.

Zhu, J., Wan, C. and Li, Y. (2010). Enhanced solid-state anaerobic digestion of corn stover by alkaline pretreatment. *Bioresour. Technol.*, **101:** 7523–7528.

Chapter 10
An Overview of the Municipal Solid Waste Management Rules in India

V. R. Sankar Cheela, Uday Shankar, and Brajesh K. Dubey

1 Introduction

Solid waste means any garbage or refuse, sludge from a wastewater treatment plant, water supply treatment plant, or air pollution control facility and other discarded material, resulting from industrial, commercial, mining, and agricultural operations, and from community activities. Nearly everything we do leaves behind waste" as per Resource Conservation and Recovery Act, USEPA, 1976. It is important to note that the definition of solid waste is not limited to wastes that are physically solid. Many solid wastes are liquid, semi-solid, or contained gaseous material (Hazra and Goel 2009).

Solid waste management (SWM) means the collecting, treating, and disposing of unwanted solid material. If not disposed or managed properly solid waste can create acute environmental and health hazards. Throughout the globe rapid urbanization is increasingly generating massive amount of municipal solid waste (MSW) the management of which poses intricate technical challenges. Also, the social and economic impacts of this phenomenon can hardly be overlooked. India is one of the swiftly growing economies in the world. Industrialization and population growth have allured people to migrate from villages to cities. As a result, thousands of tons of MSW are being generated daily in most of the cities. This MSW is expected to rise considerably by 2020 as India maintains its industrial and economic development (Shekdar 1999).

V. R. S. Cheela · B. K. Dubey (✉)
Environmental Engineering and Management Division, Civil Engineering Department, Indian Institute of Technology Kharagpur, Kharagpur, West Bengal, India
e-mail: bkdubey@civil.iitkgp.ac.in

U. Shankar
Rajiv Gandhi School of Intellectual Property and Law, Indian Institute of Technology Kharagpur, Kharagpur, West Bengal, India

D. Sengupta et al. (eds.), *Treatment and Disposal of Solid and Hazardous Wastes*,
https://doi.org/10.1007/978-3-030-29643-8_10

According to Govt. of India, 62 million tons of waste is generated annually in the country at present, out of which 5.6 million tons is plastic waste, 0.17 million tons is biomedical waste, hazardous waste generation is 7.90 million tons per annum and 15 lakh tons is e-waste. It is stated that only about 75–80% of the municipal waste gets collected and only 22–28% of this waste is processed and treated (Cheela and Dubey 2019; PIB 2016).

Environment Protection Act, 1986 confers necessary power upon the government to notify rules regarding solid waste management. The Municipal Solid Wastes (Management and Handling) Rules provides for enabling environment to guarantee methodological disposal of the solid waste generated in dwelling areas. Much of the success on the implementation depends upon multiple factors related with the management plan. The restructuring of the rules on regular basis may not be of immense significance unless the rules are examined in the light of challenges confronted in implementation stage. The nature of challenges could be either technical or regulatory. Therefore, it is incumbent to undertake a study on to contextualize the challenges in the enforcement of the Rules and suggest measures to address the identified challenges.

2 Review of Technical and Social Aspects of Solid Waste Management Rules

It is generally regarded that waste management is the sole duty and responsibility of local authorities, and that the public is not expected to contribute (Vidanaarachchi et al. 2006). Also, the engagement of stakeholders is equally important to bring operation efficiency of the waste management. The decision making process should involve both municipality and other stakeholders such as citizens and private entities (Sharholy et al. 2008). As the generation of solid waste is very domesticated activity, it is also needed to create awareness amongst community members and design possible solutions through broad-based engagement of all affected parties (Moghadam et al. 2009).

The source to meet the expenditure required for carrying out solid waste management is always a matter of concern for the municipalities. Municipalities, generally, fail to generate revenue required to create infrastructure required for giving the services. Municipalities have failed to manage solid waste due to financial factors (Sharholy et al. 2007). Sharholy et al. (2008) indicated that the involvement of the private sector is a factor that could improve the efficiency of the system. Though in India, there is a constitutional provision to allocate revenue to municipalities for carrying out their responsibilities entrusted to them by the 73rd amendment of the Constitution, the financial concern continues to be one of the major bottlenecks in effective discharge of the constitutional responsibilities.

Universities, research centres and centres of excellence have a very important role in preparing professionals and technicians in environmental fields, including waste management. Some developing countries have already seen the positive effects of investing in education and research by having cleaner cities, citizens assuming their responsibilities and higher status of solid waste workers (Guerrero et al. 2013). The lack of financial resources, inefficient institutional arrangement, inappropriate technology, weak legislative measures and unawareness in public towards solid waste management has made the service most unsatisfactory and inefficient. India is a vast country divided into different climatic zones, different food habits, and different living standard thereby producing waste of different types. Till date, no comprehensive study has been conducted to cover almost all cities and towns of India to characterize the waste generated and disposed on landfill. The policy-makers rely on the limited source of information available from few places; thereby are unable to provide appropriate solutions for the kind of waste produced for a region. There is a need to create dedicated group of officers and skilled staff for ULBs with specialization in MSWM. Adequate training and hands-on experiments would enable them to identify bottlenecks at implementation level and take appropriate action (Aditee et al. 2016; Atiq et al. 2016; Joshi and Ahmed 2016; Neha et al. 2015).

3 History of Solid Waste Management (SWM) Policies and Initiatives in India

During 1960s, the Ministry of Food and Agriculture (MoFA), Government of India declared loans for composting of solid waste. In line with the policy developments in waste management across the globe, India has developed Environmental Protect Act during the year 1986. Further, in the year 1989, hazardous waste management rules were drafted. In the year 1994, after the outbreak of plague in Surat, the policy initiatives in the field of SWM gained momentum in India. The planning commission has constituted a committee under the leadership of J.S. Bajaj in the year 1995. The recommendations include segregation of waste at source, collection of waste, composting and landfilling. A National Mission on Environmental Health and Sanitation was initiated by Ministry of Health and Family Welfare. During the year 1998, committee was constituted under the leadership of Barman to draft a policy for the solid waste management in India. The policies developed were notified in September 2000, as Municipal Solid Waste (Management and Handling) rules by Ministry of Environment, Forests and Climate Change (MoEFCC). Parallel to this, draft policy paper on funding, operation manual and requirements for MSWM were developed by Central Public Health and Environmental Engineering Organization (CPHEEO) under MoUD (Fig. 10.1).

Central Pollution Control Board (CPCB) and the State Pollution Control Boards (SPCBs) were identified as the nodal agencies to monitor the implementation of the

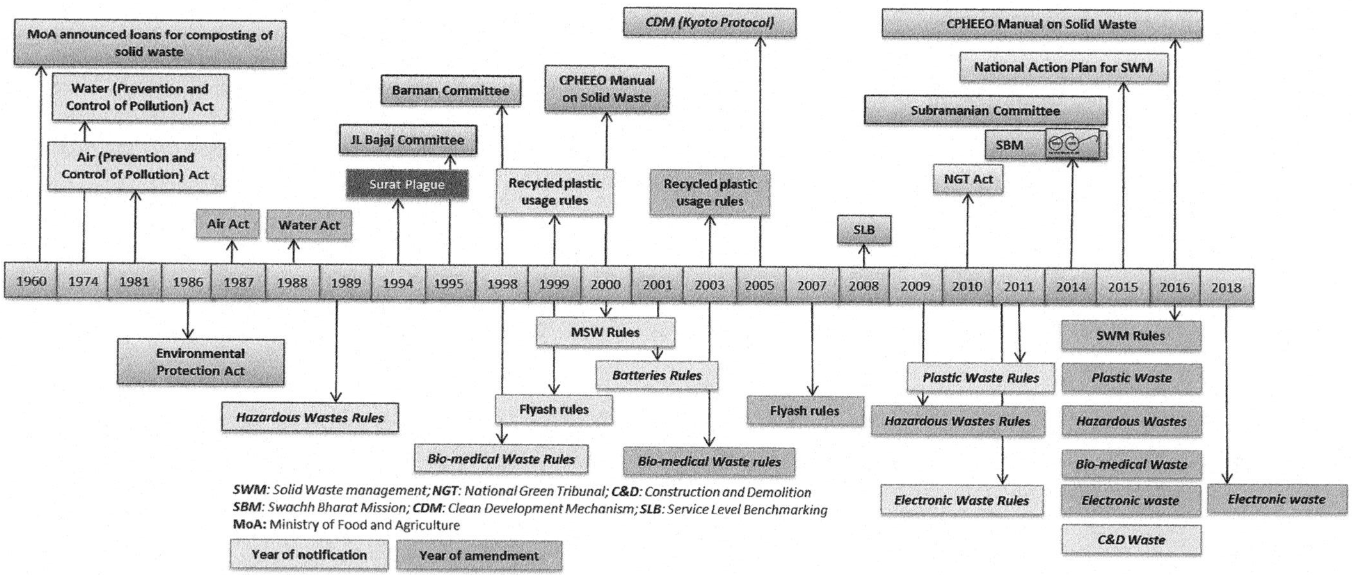

Fig. 10.1 Timeline chart of solid waste management activities in India

waste management policies and regulations across the nation. Clean development missions were initiated across the country in line with the recommendations of the Kyoto protocol in 2005. Further, MoUD has developed a service level benchmarking process to evaluate the progress in the implementation of the waste management services. In the year 2010, National Green Tribunal (NGT) was established under NGT Act, 2010 to mentor, monitor and modulate the environmental protection and conservation of forests and other natural resources. In 2014, a flagship programme namely Swachh Bharat Mission (Clean India Mission) was launched to develop scientific waste management systems and to create awareness among various stakeholders across the nation. Further, MoUD has constituted a high-steered committee under the leadership of Subramaniam to develop recommendation by introspecting the waste management systems being implemented during the past 14 years. Further based on the outcomes of the study, the government has revised the waste management rules during the year 2015. The rules were notified in the year 2016 and termed as Solid Waste Management Rules, 2016 (CPHEEO 2016; CPCB 2015).

4 Overview of Solid Waste Management Rules 2016

India had its laws in the form of the Municipal Solid Wastes (Management and Handling) Rules, 2000 notified under Environment (Protection) Act, 1986. The notification of these Rules was largely the outcome of two landmark decisions of the Supreme Court[1]. Under the Rules major responsibility was given to urban local bodies (ULBs). It was an obligatory function of ULBs to plan and manage solid waste. But in practice, it was given less priority and not much progress was made in terms of collection, segregation, transportation and suitable disposal of waste across the cities and towns. It was also observed that municipalities were short of funds and lack technical capability for proper waste management. Therefore, the need was felt to introduce community-based system where communities would be given primary responsibility to collect, sort and recycle waste.

The new 2016 Rules are aimed to rectify some of the shortcomings of the earlier regime. These Rules are again the part of extended protection umbrella of EPA, 1986 along with five other significant rules. Those are plastic, e-waste, biomedical, hazardous and construction and demolition waste management rules. Apart from municipal areas, 2016 rules are extended to census towns, urban agglomerations, areas that come under the control of Indian Railways, notified industrial townships, airports, SEZs, places of pilgrimage, religious and historical significance and State and Central Government organizations. The Rules particularly focus on segregation at source, collection and disposal of sanitary waste, imposition of responsibility on

---

[1]Dr. B. L. Wadhera v. Union of India, AIR 1996 SC 2969; Almitra H. Patel v. Union of India (1998) 2 SCC 416.

the brand owners to pull back their packaging waste, user fees for collection. Some other important changes that the new Rules bring are:

- Developing a formal integrated sector for rag pickers, waste pickers and kabadiwalas by the state government who so far work on their own.
- Absolute prohibition on throwing, burning or burying the solid waste on streets, drains and water bodies or open public spaces.
- Processing, treatment and disposal of bio-degradable waste through composting or bio-methanation within the premises as far as possible. Excess waste shall be handed over to the waste collectors or agency as controlled by the local authority.
- For SEZs, industrial estate, industrial park at least 5% of the entire area of the plot or minimum five plots/sheds are to be kept for recovery and recycling facility.
- For local bodies having a population of one million or more, waste processing facilities should be set up within 2 years.
- For census towns having population less than one million or for all local bodies with a population of 0.5 million or more, common, or stand-alone sanitary landfills should be set up within 3 years' time.
- Along with these, common or regional sanitary landfill facilities are also to be developed by all local bodies and census towns having a population under 0.5 million. This task must be completed within 3 years.
- The Rules also contain provisions for bio-remediation or capping of old and abandoned dump sites in 5 years' time.
- The Rules promote composting and ensure promotion of co-marketing of compost.
- The Rules mandate that the landfill site should be located at 100 metres distance from a river, 200 metres from a pond, 200 metres away from highways, habitations, public parks and water supply wells and 20 km away from airports/airbase. However, no landfill facility should be constructed on hilly areas. Such facility should be identified in the plain areas, within 25 kilometres. Although, transfer stations and processing facilities may be active in the hilly areas.
- All fuel-based industries located within 100 km from a solid waste-based Refuse-Derived Fuel (RDF) plant are required to decide within 6 months to substitute at least 5% of their fuel requirement by RDF so produced.
- The Rules further provides that the non-recyclable waste with calorific value of 1500 kcal/kg or more shall be used for producing energy either through RDF not disposed of on landfills or through refuse derived fuel or by giving away as feed stock for preparing refuse derived fuel. Wastes having high calorific value should be utilized for co-processing in cement or thermal power plants.
- It should be the responsibility of the Ministry of New and Renewable Energy Sources to facilitate infrastructure creation for waste to energy plants and provide appropriate subsidy or incentives to such units.
- Further, to observe the implementation of the Rules, the government has constituted a Central Monitoring Committee under the chairmanship of Secretary, MoEF&CC.

5 Challenges in Implementation of SWM Rules 2016

The SWM rules 2016 provided details for better management of solid waste. The following are the points to be addressed for better implementation of rules by the stakeholders. Vide G.S.R 451 (E) of Part II – Sec 3(ii) Part 4.1(a) duties of the generators, source segregation is mandated to channelize the waste wealth. Segregation of waste into three streams is introduced. The literacy rate of India is 65% and socio-economic status being below GDP level. Hence, constant measures should be implemented with proper base line studies before implementation of the rules as per the time-frame provided involves physical, social, economic, legal and technical aspects.

- Base line studies are to be conducted to understand the existing practices in each locality. The emphasis should be given on socio-economic characteristics of the stakeholders involved in the project; infrastructure facilities in terms of human and economic resources; awareness programmes on roles and responsibilities; capacity building; adequate support facilities to address the technical issues in implementation of technologies; regular monitoring and maintenance of records by the ULB's; and implementation of appropriate technologies based on the local conditions.
- In the section 4 of the gazette the roles and responsibilities of the generators were indicated. To make them instrumental technical facility centres, counters at local bodies, constant awareness and education programmes, incentives and rewards for best practices, subsidies for starts, provision of physical and manual infrastructure, and frame-work for collection systems should be developed.
- In the sections 5–10 of the gazette duties of the ministries were formulated. Formulation of Central monitoring committee which reviews the yearly progress has been discussed. As per the duties assigned for Ministry of Urban Development, guidelines are to be developed for formulation of national, state level policy and strategies, promotion of research and development, capacity building, technical guidance, periodic reviews on measures taken by local bodies have been assigned, with interlinking of external agencies for funding.
- The SWM rules mandate setting up of waste processing facilities by all the local bodies having population equal to or more than one lakh within a span of 2 years; Development of sanitary landfill for all local bodies having 0.5 million or more population in a period of 3 years; and Bio-remediation or capping of old and abandoned dump sites in 5-year period. Documented data of waste characteristics, quantities, availability of land, sound knowledge on technical aspects, skilled workers, adequate infrastructure and socio-economic behaviour of stakeholders play vital role in reaching the above goals.
- A framework including the policies should be developed to convert rag pickers as componential feeder source for treatment units or industries.
- Setting up of waste to energy plants, compost units and engineered landfills requires a comprehensive study on the waste composition. The climatic

conditions, socio-cultural behaviours, amount of raw material, technical and skilled human resources play a pivotal role.
- As a part of corporate social responsibility, brand users and manufacturers of products can do awareness programmes. As the accessibility of manufacturers to public is an indirect approach, framework must be developed to involve public private partnerships to build up good educational programmes.

References

Aditee, P., Anju, S., Seema, U., Neelima, N., Mayuri, N. and Indrayani, N. (2016). Innovation in solid waste management through Clean Development Mechanism in India and other countries. *Process Safety and Environmental Protection*, **101:** 160–169.

Atiq, U.Z., Shahidul, M. and Hasan, S. (2016). Performance evaluation and benchmarking of global waste management systems. *Resources, Conservation and Recycling*, **114:** 32–41.

B.L. Wadhera v. Union of India, AIR 1996 SC 2969; Almitra H. Patel v. union of India (1998) 2 SSC 416.

Cheela, V.R.S. and Dubey, B. (2019). Review of Application of Systems Engineering Approaches in Development of Integrated Solid Waste Management for a Smart City. *In:* Rathinasamy, M. et al. (eds), Water Resources and Environmental Engineering II. Springer. pp. 159–177.

CPCB (2015). Report on "The National Action Plan for Municipal Solid Waste Management" dated 5th February 2015.

CPHEEO (2016). Report on "Manual on Municipal Solid Waste Management" http://cpheeo.nic.in/SolidWasteManagement2016.htm.

Guerrero, L.A., Maas, G. and Hogland, W. (2013). Solid waste management challenges for cities in developing countries. *Waste Management*, **33:** 220–232.

Hazra, T. and Goel, S. (2009). Solid waste management in Kolkata, India: Practices and challenges. *Waste Management*, **29:** 470–478.

Joshi, R. and Ahmed, S. (2016). Status and challenges of municipal solid waste management in India: A review. *Cogent Environmental Science*, **2:** 1–18.

Moghadam, M.R.A., Mokhtarani, N. and Mokhtarani, B. (2009). Municipal solid waste management in Rasht City, Iran. *Waste Management*, **29:** 485–489.

Neha, G., Krishna, K.Y. and Vinit, K. (2015). A review on current status of municipal solid waste management in India. *Journal of Environmental Sciences*, **37:** 206–217.

PIB (2016). Report on "Solid waste management rules revised after 16 years: Rules now extend to urban and industrial areas" dated April 5, 2016.

Sharholy, M., Ahmad, K., Mahmood, G. and Trivedi, R.D. (2008). Municipal solid waste management in Indian cities – A review. *Waste Management*, **28:** 459–467.

Sharholy, M., Ahmad, K., Vaishya, R.C. and Gupta, R.D. (2007). Municipal Solid Waste Characteristics and Management in Allahabad, India. *Journal of Waste Management*, **27:** 490–496.

Shekdar, A.V. (1999). Municipal Solid Waste Management: The Indian Perspective. *Journal of Indian Association for Environmental Management*, **26(2):** 100–108.

Vidanaarachchi, C.K., Yuen, S.T.S. and Pilapitiya, S. (2006). Municipal solid waste management in the Southern Province of Sri Lanka: Problems, issues and challenges. *Journal of Waste Management*, **26:** 920–930.

Chapter 11
Environmental Implications of High Radiation in Beach Placers

Shayantani Ghosal and Debashish Sengupta

1 Introduction

One of the most attractive types of tourist destinations around the world is the beaches. The calming and soothing sound of the waves and the feeling of the silky sand at the feet attract millions of people to these destinations. The said silky sand which is a distinguishing feature of a beach, if not the most important one, is the topic of our discussion for this chapter. A number of important heavy minerals like illmenite, rutile, garnet, zircon, monazite and even metals like gold, cassiterite and platinum group of metals and even gem stones are economically exploited from these beaches. The assemblage of valuable minerals formed by gravity separation during the formation of sedimentary strata is termed as placer deposit. The beach areas which have a high concentration of these heavy minerals are known as beach placers. The Geelwal Karoo coastal placer deposits of South Africa are noted for its illmenite and garnet deposits (Macdonald and Rozendaal 1995). The Namakwa beach placer deposit of the west coast of South Africa is also noted for its zircon and gem stone deposits (Philander and Rozendaal 2015; Rozendaal and Philander 2000). In Brazil the Guarapari beach area is a popular tourist spot which is noted for its high radioactive dose rate values. According to Vasconcelos et al. (2013), the radioactive dose rate values of the area is almost 187 times higher than the prescribed limit by UNSCEAR (2000). This is mostly because of the presence of high concentration of the element thorium which is present in the mineral monazite. Monazite is a thorium phosphate mineral, which consists of 70% of Rare Earth Element oxides, 4–6% of thorium oxide and 0.15–0.25% of uranium oxide (Malarica et al. 1994). Apart from Guarapari several other beach areas of Brazil have a radioactive absorbed dose rate that is higher than the limit prescribed by the United Nations Scientific

S. Ghosal · D. Sengupta (✉)
Department of Geology and Geophysics, Indian Institute of Technology Kharagpur, Kharagpur, West Bengal, India
e-mail: dsgg@gg.iitkgp.ac.in

D. Sengupta et al. (eds.), *Treatment and Disposal of Solid and Hazardous Wastes*,
https://doi.org/10.1007/978-3-030-29643-8_11

Committee on the Effects of Atomic Radiation (2000). Certain beach areas of Rio de Janeiro like Mambucaba and Buena; Meaipe and Anchieta areas of Espirito Santo (Veiga et al. 2006) have a high concentration of thorium which thereby results in a high radioactive dose rate values. The high thorium concentration has not only affected the beaches of Brazil but certain other beaches around the world as well. The Potenga beach areas of Chittagong, Bangladesh also shows a high concentration of thorium (Yasmin et al. 2018). The Miami Bay, Penang Island beach areas of Malaysia has monazite-rich black sand deposits and these areas show a high concentration of thorium and radium (Shuaibu et al. 2017). High concentration of thorium is also observed in certain samples of Hadhramout coast, Gulf of Aden, Yemen (Badran et al. 2016).

Almost 85% of the radiation that affects living population comes from the natural radionuclides (UNSCEAR 2000). Beach sands along with their heavy mineral content consist of these radionuclides, the proportion of which depends on the local geology of the beach area. Moreover beach sands are used for industrial and construction purposes, so natural radioactivity due to the presence of ^{238}U, ^{232}Th and ^{40}K should be measured. India has the second largest thorium deposits of the world and most of these deposits are present along the coastal regions of the country. In India the black beach sands of Kerala (UNSCEAR 2000) are already known for its high radioelement concentration. Along the south east coast of Tamil Nadu certain beach areas show a high concentration of thorium followed by uranium and potassium. In fact a close look along the eastern coastal region of the country shows that a high radioelement concentration, particularly a high thorium concentration, is prevalent along the beach areas. Apart from the beach areas of Tamil Nadu, the Bhimlipatnam and Ramakrishna beach areas of the state of Andhra Pradesh also show a high thorium concentration (Palaparthi et al. 2017). The beach areas of the state of Odisha have been extensively studied for its heavy mineral concentration and also to find the radioelement concentration. The beach areas of Ersama, Chhatrapur, Gopalpur, Garampeta, Markandi and also areas close to the mouth of the river Rushikulya show a high radioelemental concentration (Mohanty et al. 2003, 2004a, b; Rao et al. 2009; Ghosal et al. 2017). It is the high concentration of the element thorium which is the main reason behind the elevated radioelement concentration.

The primary reason behind this elevated thorium concentration is the surrounding geology of this area. The eastern side of the Indian subcontinent consists of a geological domain known as the Eastern Ghat Mobile Belt (EGMB) which consists of typical rock types: charnockite, khondalite, granite, migmatite and anorthosite (Ramakrishnan et al. 1998). These rock types particularly charnockite and khondalite consist of monazite and zircon as accessory mineral. The presence of thorium in monazite and uranium in zircon is the primary source of radioelements in the beach area. The constant weathering of these rock types of the EGMB by the surrounding river system is resulting in the transportation of the sediments to the beach area. The wave and tidal activities of the ocean results in the distribution of these sediments along the length of the beach. It is the presence of these sediments that results in the enrichment of the beach sands in heavy minerals containing monazite and zircon resulting in an elevated background radiation of the area.

Here we are going to talk about the High Background Radiation Areas (HBRA) along the eastern coastal region of the Indian subcontinent particularly the coastal areas of the state of Odisha. Studies conducted along certain beach areas of Odisha show a high background radiation value. For this purpose we are studying five beach areas (Ersama, Rushikulya, Chhatrapur, Garampeta and Markandi beaches) of Odisha and three beach areas of Andhra Pradesh (Sivasagar, Bhimlipatnam and Ramakrisha beach). The general objectives are as follows:

1. The assessment and concentration of radionuclides (U, Th and K) in the beach sands. It will help us to understand the subsequent environmental hazard associated with it.
2. The improper management of these beach sands is leading to the usage of these sands in construction purposes which is exposing a greater number of populations to the radiation associated with these beach sands. Hence, the radiological impact is calculated by computation of the radioactive dose rate.
3. A comparison between the radioactive data of the beach areas of Odisha and Andhra Pradesh has been discussed in order to find out the radiological potential of these areas.
4. Radium equivalent and Hazard Index estimation of these beach areas has been done.

2 Materials and Methods

Sand samples were collected along various beach areas of Odisha and Andhra Pradesh (Fig. 11.1). The samples were collected maintaining a northeast southwest trend. At each site the samples have been collected maintaining a gap between 500 m and 1 km. During collection, the top portion of the sand has been removed to avoid collecting rootlets and other anthropogenic objects. These samples collected from a shallow depth is dried and sieved through 0.355 mm in the laboratory. Then these samples are stored in glass air-tight vials and sealed to check for radon escape. The samples are then rested for a month to attain secular equilibrium.

2.1 Radiometric Analysis

The natural radioactivity of an area depends on a number of radionuclides, among them the most important ones are ^{232}Th, ^{238}U and ^{40}K. These radionuclides have been measured by γ-spectroscopic analysis. It is done in the Radiochemistry Division (BARC), Variable Energy Cyclotron Centre (VECC), Kolkata, India. Based on absolute method of analysis, activity concentration of uranium, thorium and potassium is measured. The analysis was done in GEM series High Purity Germanium (HPGe) coaxial detector, ORTEC, USA with 50% efficiency and 1.8 keV energy resolution at 1332 keV of ^{60}C. The detector efficiency has been calculated by using

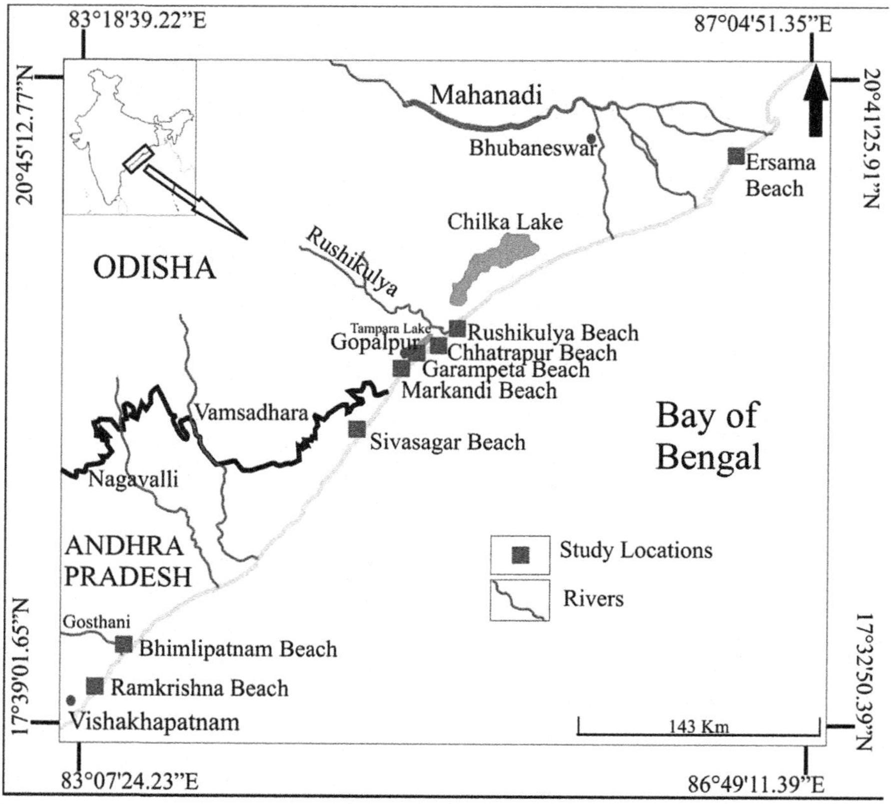

Fig. 11.1 Coastal areas of the eastern side of peninsular India. The various study areas along the beaches of Odisha and Andhra Pradesh have been marked by the square boxes. The relevant rivers and lakes have also been highlighted

^{152}Eu standard source of known activity with the help of Eq. (11.1). Potassium can be measured by its own γ-rays, but ^{232}Th and ^{238}U do not emit ϒ-rays directly. These can be measured by tracking the γ-rays emitted by their respective daughter products. Daughter products of ^{232}Th [338,911(^{228}Ac), 239(^{212}Pb), 727(^{212}Bi), 583 (^{208}Tl)] and ^{238}U [295,352(^{214}Pb) and 609(^{214}Bi)], are considered assuming a secular equilibrium status for the decay series. Each sample is measured for 40,000 s to obtain a reasonable ϒ-ray peak area. From this peak area the activity is measured with the help of a multi-channel analyser Canberra DSA 1000, which is connected to computer through Genie 2 K software.

$$\in = \frac{CPS \times 100}{A_t \times I} \quad (11.1)$$

where $\in$ = Efficiency, CPS = Counts per second, A_t = Activity of ^{152}Eu and I = Intensity of ^{152}Eu.

2.2 Absorbed Dose Rate

The radiation received by a person per kilogram of its mass is the absorbed dose rate by the person. The conversion factor required to calculate absorbed dose rate at 1 m above the ground surface, due to uniform distribution of gamma radiations from ^{238}U, ^{232}Th and ^{40}K is calculated by the method proposed by the UNSCEAR (2000). The equation is mentioned below:

$$D(\text{nGyh}) = 0.461A_{\text{U}} + 0.623A_{\text{Th}} + 0.0414A_{\text{K}} \tag{11.2}$$

where A_{U}, A_{Th} and A_{K} are specific activity of ^{238}U, ^{232}Th and ^{40}K in Bq kg^{-1}.

2.3 Annual Effective Dose Rate

Annual effective dose rate gives us an idea about the internal and external radiation exposure. It is calculated using the conversion coefficients (0.7 SvGy^{-1}) and 0.2 as the outdoor occupancy factor, as proposed by UNSCEAR (2000).

$$\text{AED}\left(\text{mSvy}^{-1}\right) = D\left(\text{nGyh}^{-1}\right) \times 8760\left(\text{hy}^{-1}\right) \times 0.2 \times 0.7\left(\text{Sv Gy}^{-1}\right) \times 10^{-6} \tag{11.3}$$

2.4 Radium Equivalent

Radium equivalent describes the gamma output from different mix of U, Th and K in samples and helps in comparison of radiation exposure due to different radioisotopes.

$$\text{Ra}_{\text{eq}} = A_{\text{U}} + 1.43A_{\text{Th}} + 0.077A_{\text{K}} \tag{11.4}$$

The equation is proposed by UNSCEAR (2000), based on the assumption that contributions from other radionuclides are insignificant.

2.5 Hazard Index

It is the index that indicates the external exposure due to the radioisotopes in the environment.

$$H_{ex} = A_U/370 + A_{Th}/259 + A_K/4810 \quad (11.5)$$

$H_{ex} < 1$ is considered to be within limit, and $H_{ex} > 1$ is unsafe and equivalent to 370 Bq kg^{-1}.

3 Results and Discussion

All of these beaches show a high concentration of black particles in the beach sand and these particles appear in patches. The concentration of the individual radioelements of the area shows that the concentration of thorium is higher followed by potassium and uranium (Fig. 11.2). The average concentration of thorium is highest in the Ersama beach areas with values ranging between 900 and 4700 Bq/kg with an average of 2825 Bq/kg. Followed by Ersama the thorium concentration shows a decreasing trend towards the southwest direction along the coastal area. The average

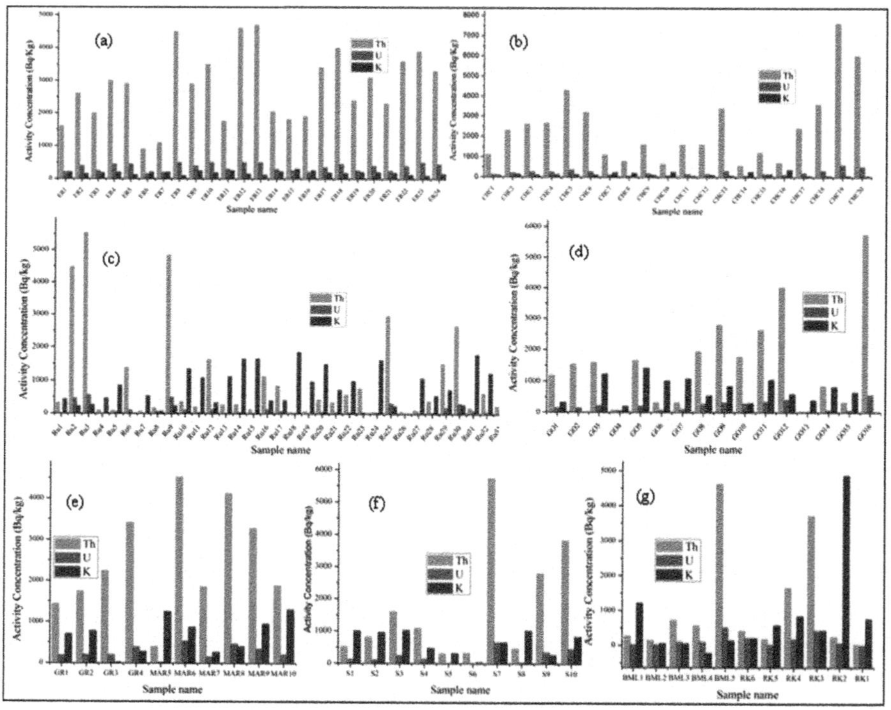

Fig. 11.2 Varying activity concentration of ^{232}Th, ^{238}U and ^{40}K along the eastern coastal beach areas of Odisha and Andhra Pradesh. Activity concentration of particular beaches are represented here: (**a**) Ersama Beach (Mohanty et al. 2004a); (**b**) Chhatrapur Beach (Mohanty et al. 2004b); (**c**) Rushikulya Beach (Rao et al. 2009); (**d**) Gopalpur Beach (Rao et al. 2009); (**e**) Garampeta and Markandi Beach (Ghosal et al. 2017); (**f**) Sivasagar Beach (Ghosal and Sengupta 2018); (**g**) Bhimlipatnam and Ramakrishna Beach (Palaparthi et al. 2017)

concentrations of thorium following the NE-SW trend along the beach areas of Chhatrapur, Rushikulya, Gopalpur, Garampeta, Markandi, Sivasagar, Bhimlipatnam and Ramakrishna beach are 2447, 1672, 991, 1845, 3133, 1745, 1263 and 1036 Bq/kg respectively. The average uranium concentrations of the study area along the beaches of Ersama, Chhatrapur, Rushikulya, Gopalpur, Garampeta, Markandi, Sivasagar, Bhimlipatnam and Ramakrishna are 358, 220, 201, 125, 204, 343, 211, 146 and 146 Bq/kg respectively. The activity concentrations of potassium from NE to SW along the Ersama, Chhatrapur, Rushikulya, Gopalpur, Garampeta, Markandi, Sivasagar, Bhimlipatnam and Ramakrishna beaches are 182, 130, 694, 854, 609, 756, 655, 242 and 1275 Bq/kg respectively. According to UNSCEAR (2000), the average ambient ^{232}Th, ^{238}U and ^{40}K values are 36, 33 and 474 Bq/kg. Almost all of the above mentioned beach areas show an average value of ^{232}Th, ^{238}U and ^{40}K greater than the permissible limit.

The average absorbed dose rate of the area along Ersama, Chhatrapur, Rushikulya, Gopalpur, Garampeta, Markandi, Sivasagar, Bhimlipatnam and Ramakrishna beach is 1932, 1631, 1161, 696, 1269, 2142, 1457, 864 and 766 nGy h^{-1}. The values are much higher than the prescribed limit of 55 nGy h^{-1}. The Gopalpur and Bhimlipatnam beach areas which are popular tourist spots also show an absorbed dose rate almost 21 and 16 times higher than the limiting value. The Annual effective dose rate value along NE to SW ranges between 4.9 and 0.03 mSvy^{-1} and the limiting value as proposed by UNSCEAR (2000) is 0.07 mSvy^{-1}. The average Radium Equivalent values of the beach area along Ersama, Chhatrapur, Rushikulya, Gopalpur, Garampeta, Markandi, Sivasagar, Bhimlipatnam and Ramakrishna beach are 4412, 3730, 1579, 2642, 2890, 4882, 2758, 1971 and 1727. Almost all of the Hazard Index values of the study area lie above the limiting value of 1 (Fig. 11.3a) which is the limit prescribed by UNSCEAR (2000).

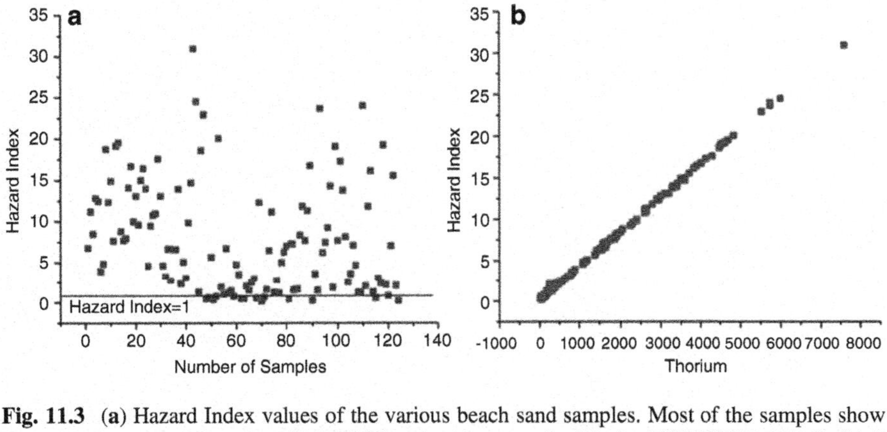

Fig. 11.3 (**a**) Hazard Index values of the various beach sand samples. Most of the samples show Hazard Index greater than one which is higher than the prescribed limit of UNSCEAR (2000). (**b**) A comparison of Hazard Index values with that of thorium concentration of the study area

From Fig. 11.1 it is observed that beach areas of Rushikulya and Gopalpur lie close to the mouth of the river Rushikulya, which is draining into the Bay of Bengal. The uranium concentration of these areas is relatively low as compared to the uranium concentration of the other beach areas (Fig. 11.2). This is because the presence of uranium in nature happens in mostly two states U^{+6} and U^{+4}; of these the +6 state is highly reactive and more prone to leaching as compared to the +4 state. In case of thorium which has only one state of Th^{+4}, it is insoluble in water (Aswathanarayana 1985).

It is evident from Fig. 11.2 that the study area has a dominant concentration of thorium when compared to uranium and potassium. The Radioactive Dose Rate and the Hazard Index of the area as observed in Figs. 11.3a and 11.4 show a high value which is harmful for the population living in the vicinity of these areas. In Fig. 11.3b a comparison between the Hazard Index values and the thorium concentration of the study area shows an absolute dependence of the hazard index values on thorium. A similar pattern is observed in case of the absorbed dose rate value (Fig. 11.4) of the area. When compared with the thorium concentration of the area it shows a positive correlation. Hence, it can be said with surety that the High Background Radiation (HBRA) values of the beach areas along the eastern coastal region of the Indian subcontinent is primarily dependent upon the thorium concentration of the study area. The presence of the mineral monazite containing thorium in its crystal lattice is the main contributing factor behind the high concentration of thorium. Also, the presence of the mineral xenotime which consists of thorium as reported by Behera (2003) and Subrahmanyam et al. (2004) along the study area can also be a contributing factor.

The illegal minning of these beach sands and their constant usage in construction purposes is an immediate threat that needs to be checked. Apart from the population that resides close to these beach areas being affected by these high radiation doses the usage of these sands in building construction is exposing a larger mass of people to the harmful effects of radiation.

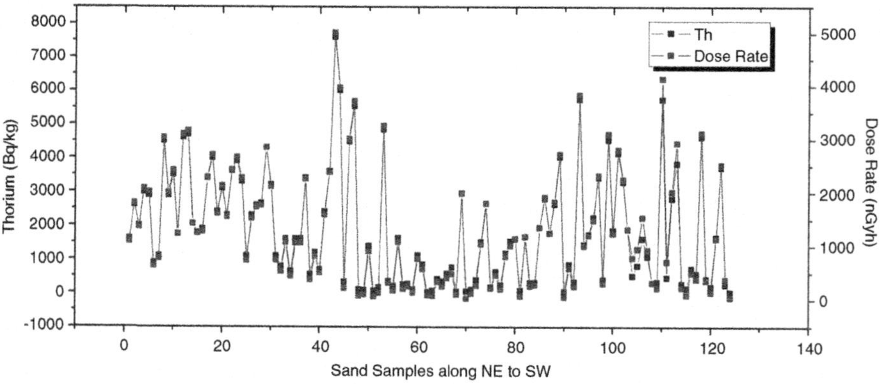

Fig. 11.4 Representation of the varying dose rates of the study area. The comparison between the absorbed dose rate and the thorium concentration of the study area shows a positive correlation

References

Aswathanarayana, U. (1985). Principles of Nuclear Geology. Oxonian Press, New Delhi, pp. 88–90.

Badran, H.M., Bin-Jaza, A.A. and Sharshar, T. (2016). Multivariate Statistical Analysis of High Background Radiation Area on the Hadhramout Coast. *Arabian Journal of Science and Engineering*, **41:** 311–323.

Behera, P. (2003). Heavy mineral in beach sands of Gopalpur and Paradeep along Odisha coast line, east coast of India. *Indian Journal of Marine Science*, **32(2):** 172–174.

Ghosal, S. and Sengupta D. (2018) Modeling of Naturally Occurring Radioactive Material in Beach Sands from Sivas Agar, Andhra Pradesh, India – A case Study. *Journal of Radiation and Nuclear Applications* 3(3), 157–161.

Ghosal, S., Agrahari, S., Guin, R. and Sengupta, D. (2017). Implications of modelled radioactivity measurements along coastal Odisha, Eastern India for heavy mineral resources. *Estuarine, Coastal and Shelf Science*, **184:** 83–89.

Macdonald, W.G. and Rozendaal, A. (1995). The Geelwal Karoo heavy mineral deposit: A modern day beach placer. *Journal of African Earth Sciences*, **21(1):** 187–200.

Malarica, A., Luce, C.N., Sogni, R., Achilli, L. and Fabbri, S. (1994). Naturally Occurring Radionuclides in the Monazite Sands of Guarapari and Meaipe (Brazil). Proceedings of the La Radioattività Ambientale nel Nuovo assetto Istituzionale Piancenza, Bologna and Ravenna, 225–230.

Mohanty A.K., Das S.K., Van K. V., Sengupta D. and Saha S.K. (2003) Radiogenic heavy minerals in Chhatrapur beach placer deposit of Orissa, south eastern coast of India. *Journal of Radioanalytical and Nuclear Chemistry*, **258(2):** 383–389.

Mohanty, A.K., Sengupta, D., Das, S.K., Vijayan, V. and Saha, S.K. (2004a). Natural radioactivity in the newly discovered high background radiation area on the eastern coast of Orissa, India. *Radiation Measurements*, **38(2):** 153–165. doi:https://doi.org/10.1016/j.radmeas.2003.08.003

Mohanty, A.K., Sengupta, D., Das, S.K., Saha, S.K. and Van, K.V. (2004b). Natural radioactivity and radiation exposure in the high background area at Chhatrapur beach placer deposit of Orissa, India. *Journal of Environmental Radioactivity*, **75:** 15–33.

Palaparthi, J., Chakrabarti, R., Banerjee, S., Guin, R., Ghosal, S., Agrahari, S. and Sengupta, D. (2017). Economically viable rare earth element deposits along beach placers of Andhra Pradesh, eastern coast of India. *Arabian Journal of Geosciences*, **10:** 201.

Philander, C. and Rozendaal, A. (2015). Detrital zircon geochemistry and U-Pb geochronology as an indicator of provenance of the Namakwa Sands heavy mineral deposit, west coast of South Africa. *Sedimentary Geology*, **328:** 1–16.

Ramakrishnan, M., Nanda, J.K. and Augustine, P.F. (1998). Geological Evolution of the Proterozoic Eastern Ghats Mobile Belt. *Geological Survey of India Special Publication*, 44.

Rao, N.S., Sengupta, D., Guin, R. and Saha, S.K. (2009). Natural radioactivity measurements in beach sand along southern coast of Orissa, Eastern India. *Environmental Earth Sciences*, **59:** 593–601. doi:https://doi.org/10.1007/s12665-009-0057-x

Rozendaal, A. and Philander, C. (2000). Mineralogy of heavy mineral placers along the west coast of South Africa. *In:* Rammlmiar, D., Mederer, J., Oberthür, T., Heiman, R.B., Pentinghaus, H. (Eds), Applied Mineralogy in Research, Economy, Technology, Ecology and Culture. Proceedings of ICAM 2000, Göttingen, 417–421.

Shuaibu, H.K., Khandaker, M.U., Alrefae, T. and Bradley, D.A. (2017). Assessment of natural radioactivity and gamma-ray dose in monazite rich black Sand Beach of Penang Island, Malaysia. *Marine Pollution Bulletin*, **119(1):** 423–428.

Subrahmanyam, A.V., Desapati, T., Kumar, V.A., Deshmukh, R.D. and Viswanathan, G. (2004). Occurrence of xenotime in the Narasapur beach placers, West Godavari District, AP. *Current Science*, **87(10):** 1458–1461.

United Nations Scientific Committee on the Effects of Atomic Radiation (UNSCEAR) (2000). Sources and Effects of Ionizing Radiation. Report to General Assembly, with Scientific Annexes, United Nations, New York.

Vasconcelos, D.C., Reis, P.A.L., Pereira, C., Oliveira, A.H., Santos, T.O. and Rocha, Z. (2013). Modelling Natural Radioactivity in Sand Beaches of Guarapari, Espírito Santo State, Brazil. *World Journal of Nuclear Science and Technology*, **3:** 65–71.

Veiga, R., Sanches, N., Anjos, R.M., Macario, K., Bastos, J., Iguatemy, M., Aguiar, J.G., Santos, A. M.A., Mosquera, B., Carvalho, C., Baptista Filho, M. and Umisedo, N.K. (2006). Measurement of natural radioactivity in Brazilian beach sands. *Radiation Measurement*, **41(2):** 189–196. https://doi.org/10.1016/j.radmeas.2005.05.001.

Yasmin, S., Barua, B.S., Khandaker, M.U., Kamal, M., Rashid Md., A., Sani, S.F.A., Ahmed, H., Nikouravan, B. and Bradley, D.A. (2018). The presence of radioactive materials in soil, sand and sediment samples of Potenga sea beach area, Chittagong, Bangladesh: Geological characteristics and environmental implication. *Results in Physics*, **8:** 1268–1274.

Chapter 12
Wastes to Wealth for Bioenergy Generation

Pintu Buda, Aishwarya Srivastava, Salma Bano, Tania Dev, Ritiritao Brahma, and Rintu Banerjee

1 Introduction

Handling waste in an artistical manner and treating it for obtaining maximum value added products and byproducts are the thrust area of waste valorization and wealth generation by the technocrats of the modern era. The amount of wastes generated is directly related to the logarithmic increase in population throughout the world. This population rise has a direct correlation which may lead to upgrading industrialization, urbanization and modernization (Smith and Lee 1993; Kundu et al. 2017; Goel 2017). According to the report of World Bank in 2016, 2.01 billion tonnes (BT) of wastes is estimated to be generated which is expected to be increased upto 3.40 BT by 2050 (Kaza et al. 2018). These rise in wastes has become a serious issue and will have a serious negative impact not only on the economy but also affect the overall development of any nation (Dincer 2000; Grossman and Krueger 1995). Further, applications of efficient modern tools like GIS and remote sensing can improve waste management systems (Dutta and Goel 2017) but the continuous effort by the researchers are still continuing since few decades who are trying to reduce the cost of management system by finding out many alternative ways (Batayneh et al. 2007; Walton et al. 1998; Wyman et al. 2009; Singh et al. 2012; Chintagunta et al. 2016) such as recycling, proper management and value addition to them (Srivastava 2007). One of the favourable way of valorization of these wastes is opting for bioenergy generation, leading towards the novel concept of "wastes to wealth" (Singh et al. 2011, 2012; Jacob and Banerjee 2016; Banerjee et al. 2017). This concept not only deals with valorization of the wastes but also helps in fulfilling the increasing demand of bioenergy in the form of various fuels like bioethanol, biodiesel, biogas,

P. Buda · A. Srivastava · S. Bano · T. Dev · R. Brahma · R. Banerjee (✉)
Agricultural & Food Engineering Department, Indian Institute of Technology Kharagpur, Kharagpur, West Bengal, India
e-mail: rb@iitkgp.ac.in

D. Sengupta et al. (eds.), *Treatment and Disposal of Solid and Hazardous Wastes*,
https://doi.org/10.1007/978-3-030-29643-8_12

and biomanure. The present chapter deals with the various types of liquid and gaseous biofuels, their classification, and methods of conversion of wastes to wealth.

1.1 Classification of Wastes

The wastes can be classified in various ways as per the source, nature, degradability, etc. Depending on the source of pollutants, the wastes are categorized into point source and non-point sources. The wastes can also be classified into organic and inorganic on the basis of chemical nature and further grouped into biodegradable and non-biodegradable. The classification of wastes has been shown in Fig. 12.1.

According to Environmental Protection Agency (EPA), U.S.A., point source pollutants are defined as pollutants discharged from any single identifiable source such as effluents discharged from wastewater treatment plants, operational wastes from industries, sewage leakages, etc. On the other hand, non-point source pollutants are those lacking any specific source and originates from a widespread region such as agricultural fields, dumping grounds, etc.

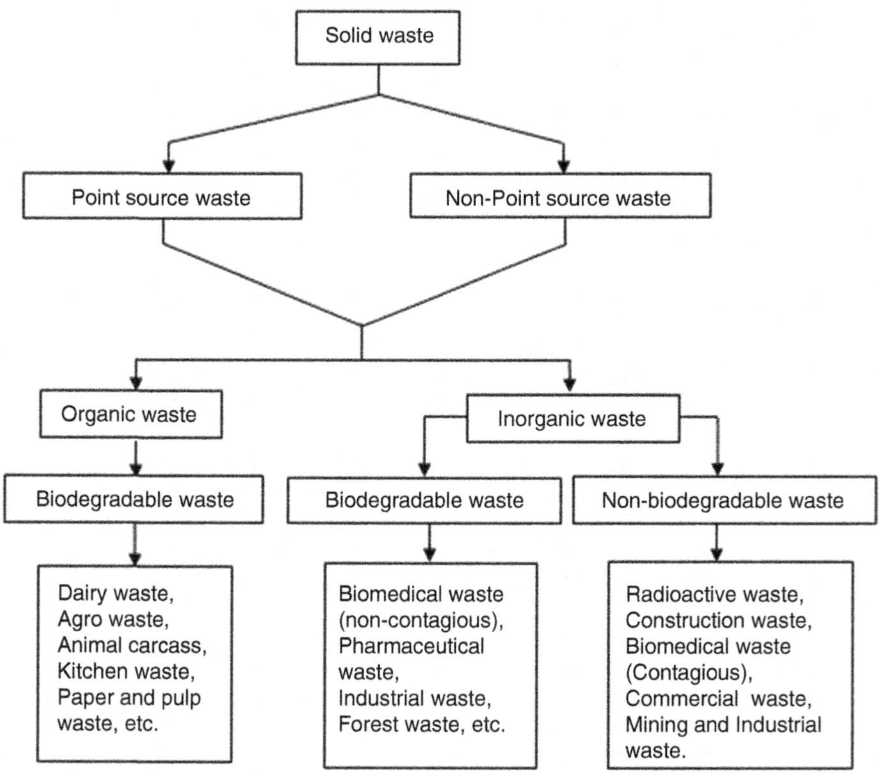

Fig. 12.1 Classification of wastes

Table 12.1 Type, origin and constitution of different wastes

Type	Origin	Constitution
Municipal solid wastes	Household	Edible wastes (vegetable and fruit peels, leftover food, wastes oil, etc.)
		Non-edible wastes (stationaries, clothes, hide wastes, etc.)
	Municipality corporation	Pruning, sludge, construction site wastes, cleaning and sweeping streets, etc.
Biomedical wastes	Contagious wastes	Gauge, gloves, cultures, contaminants, blood and bodily fluids, swabs, etc.
	Hazardous wastes	Surgicals, disposables and chemicals, etc.
Industrial wastes	Food sector	Packaging wastes, food wastes, whey, sugar, etc.
	Mining sector	Coal ash, blast and steel furnace slag, brine mud, kiln dust, phosphogypsum, red mud, iron tailing, etc.
	E-wastes	Electronics packaging, discarded computers and other electronic gadgets, cartridges, wires, electrical boards, etc.
Agro-forest wastes		Spoiled food wastes, husk of grains, fodders, trunk, branches, fallen leaves, etc.
Institutional wastes		Paper, cardboards, plastics, computer peripherals, etc.

Organic wastes are generally derived from living beings, like the human, animal or plant life. These are easily bio-degraded either naturally or by help of certain mediators like enzymes, microbes, physical conditions, chemicals, etc. into smaller fragments which can be absorbed by the soil and water and can be returned to the nature through the biogeochemical cycles. Inorganic wastes coming from industrial processes and non-renewal natural resources like mineral and petroleum, include materials that are difficult to biodegrade as well as easily biodegradable depending on the origin (Table 12.1).

The wastes can be further classified into biodegradable and non-biodegradable based on the sources from which they are generated. Biodegradable wastes such as lignocellulosic wastes viz., vegetable wastes, agricultural wastes, solid food, paper industry wastes, etc. (Das and Singh 2004; Davis and Song 2006; Demirbas 2011; Singh et al. 2012; Avanthi et al. 2017) can be readily segregated from non-biodegradable wastes such as plastic and/or metals-based wastes (Demirbas 2011; Zheng et al. 2005). The major types of wastes have already been discussed in Fig. 12.1 and the types of waste generated along with its constitution is tabulated in Table 12.1.

1.2 Characterization of Wastes

Characterization acts as one of the most prudential step in recycling of wastes for its valorization aiding the downstream process configuration (Goel 2008; Kumar and

Goel 2009). It involves all the studies related to physical, chemical and biological properties of the wastes which help in deciding the value-added product generation (Singh and Chandel 2018; Goel 2017). Depending upon the nature of origin, the governing parameters such as chemical oxygen demand (COD), biological oxygen demand (BOD), and carbon content differ amongst the wastes (Kandakatla et al. 2013; Khan et al. 2017). The physical, chemical and biological characterization further helps in ameliorating wastes segregation based on their biodegradability and thereby, reduces cost of the primary product through circular economy at industrial level (Singh et al. 2011; Singh et al. 2012; Kandakatla et al. 2013; Goel 2017). Likewise, value addition of the wastes generated from domestics, hubs, offices, etc. can be designed. The difficulty in segregation is directly affected by increased number in sources of wastes generations at various environmental conditions. This can be addressed on proper characterization of wastes.

Physical, Chemical and Biological Characterization of Wastes
Physical parameters such as pH, density, viscosity, particle nature and size, moisture content, etc. are essential to be evaluated to know the nature of the waste (Demirbas 2011). In the next step, chemical properties, which include biochemical composition, elemental composition, biological oxygen demand (BOD), chemical oxygen demand (COD), metal ion content, etc. aid in the process configuration. Sometimes, the micro-flora and fauna in the wastes may interfere positively or negatively in the value addition process (Olmstead and Weber Jr. 1991; Gupta and Mohapatra 2003). One can optimize the process configuration (reactor design, microbial system, temperature, pH etc.) to enhance the efficiency of overall top-down process which are essential benchmark for appropriate design (Mor et al. 2006; Jacob and Banerjee 2016). While characterizing the physical parameters, some important factors such as heavy metals (monovalent and divalent anions/cations), insecticides, pesticides, herbicides, etc. are playing a significant role for taking the appropriate decision on treatment process. Similarly, the importance of microbial load such as presence of bacteria, mold, fungi, actinomycetes, viruses, protozoa, porifera, etc. are also playing a significant role whose presence and concentrations are essential for making suitable reactor design. At the same time, total solids, dissolved solids, suspended solids, volatile solids and dissolved oxygen are also essential parameter for deciding the treatment method of any waste.

2 Pretreatment Technologies for Wastes

Pretreatment of characterized biomass has become an intangible step towards biofuel production. Recently use of anaerobic pretreatment methods have superseded aerobic methods owing to lower environmental adversity and cost effectiveness (Mosier et al. 2005). A consolidated explanation on physical, physico-chemical, chemical and biological pretreatment of wastes along with its advantages and disadvantages on selected process parameters has been tabulated in Table 12.2.

Table 12.2 Advantages and disadvantages of different pretreatment processes

Method	Technique	Advantages	Disadvantages
Physical pretreatment	Milling	Breakdown of cell biomass	Huge power consumption
	Pyrolysis	Rapidly hydrolyze the lignocellulosic matrix	Not effective process
Physicochemical pretreatment	Steam explosion	Depolymerization of lignin and holocelluloses	Generates toxic products
	Ammonia fiber explosion	The surface area is increased.	Not applicable for high lignin containing biomass
		No production of toxic inhibitors	
	Carbon dioxide explosion	Cost effective	Do not effectively depolymerize the lignocellulosic biomass
Chemical pretreatment	Ozonolysis	Effective process.	Costly method
		No production of toxic inhibitors	
	Acid hydrolysis	Effectively depolymerize lignin and hydrolyze sugars	Have corrosive effects.
			Produces toxic inhibitors
	Alkaline hydrolysis	Increases the surface area of the matrix	Time consuming process.
			Salts are formed and get incorporated in feedstock
	Oxidative delignification	Effectively depolymerize biomass	Cost ineffective, time and labour consuming.
Biological pretreatment	Direct microbial attack	Contamination problems	Slow process
	Enzymatic delignification	Requires low energy.	Enzyme titre and time of delignification depends on types of feedstock
		Highly specific and effective process.	
		Eco-friendly process	

3 Processing of Wastes

Treatment plant design is one of the most challenging task for the biotechnologists depending upon the initial nature of the waste. Different pretreatment methods such as primary, secondary, and tertiary treatment methods are generally adopted based on the initial characteristics of the wastes. Moreover, the amount of sludge and slurry present in the wastes is also playing a significant role. If the waste has high quantity of liquid, dewatering is one of the steps which can be achieved for concentrating the liquid wastes through centrifugation, vacuum filter, press filter, horizontal belt filter, drying bed, lagoons, etc. Although there are no standard thumb rule for such treatment processes, but the valorization of wastes can be carried out through different processes where BOD/COD and DO are playing a significant role. The valorization of wastes are very much product dependent; thus, for production of single cell protein (SCP), citric acid, succinic acid, levulinic acid, acetic acid, formic acid, lactic acid, bioethanol, biodiesel, biobutanol, xylitol, etc. can be produced with appropriate selection of aerobic and anaerobic processes. In the modern era, the selection of both aerobic/anaerobic process in the beginning of the treatment method is also getting popularized; thus, the entire responsibility for appropriate treatment process selection is totally dependent on the targeted end product formation.

Improvement of biomass usage requires enormous efforts to develop new biomass frameworks in which production, transformation and use of bio-based products are done productively in close agreement with nature. In the subsequent section, waste valorization utilizing different lignocellulosics have been documented for 1G and 2G ethanol production.

4 Products: Liquid Fuels and Gaseous Fuel

Biomass-based fuels are an excellent alternative to the non-renewable fossil-based fuels due to their superior qualities such as carbon neutralizing and reducing global warming. Wastes biomass could be transformed to biofuels and other useful byproducts through biomass upgrading and biorefinery technologies. Biomass upgrading process includes different physical, chemical, physico-chemical, thermal and biological processes to generate bioethanol, biobutanol, biodiesel, biogas and many more.

4.1 Bioethanol (1G and 2G)

Production of bio-based ethanol is being carried out from centuries but recent advancement in the industrialization sector has enhanced its valorization in the fuel market. Bioethanol is a substitute of petrol and it also serves as the feedstock for ethyl tertiary butyl ether. Bioethanol can easily blend with petrol and can be used in the flexi-fuel vehicles (Rajak and Banerjee 2018). Biomass consisting carbohydrates, especially holocellulose can be successfully converted into bioethanol.

4.1.1 1G Ethanol

The bioethanol production from edible agricultural feedstocks is an efficient process but comes with concerns over long term sustainability raising the food versus fuel issue. The major highlights and challenges are lying with the present generation researchers to improve the production and productivity statistics for bioethanol generation. It is a continuous process for ethanol production. The starchy-based residues undergo rigorous saccharification process where the amylose and amylopectin fractions of starchy raw materials undergo biotransformation in presence of other enzyme or whole cell to produce reducing sugars. This process may be aerobic and facultative one where low stress of oxygen may be utilized during the conversion process whereas in case of anaerobic fermentation of yeast (*Saccharomyces cerevisiae*) reducing sugar is converted to ethanol. In first generation ethanol production, the starchy biomass such as potato, sugarcane juice, cassava, etc. are the major feedstocks which do not require pretreatment and thus, time for ethanol production is shorter in comparison to the second generation process.

4.1.2 2G Ethanol

Lignocellulosic biomass accumulated from distinct origins differ in their organic composition and the major bottleneck towards biofuel production is the presence of lignin polymer and crystalline nature of cellulose. Pretreatment processes are designed to disrupt lignin structure, increase the surface area of cellulose increasing accessibility to sugar moieties without damaging monomeric sugars and producing inhibitory byproducts which can interfere with further downstream processes (Ariunbaatar et al. 2014; Kumar et al. 2009; Mosier et al. 2005). The different lignocellulosics contain different ranges of lignin (Table 12.3) where during pretreatments many intermediate products such as phenolic compounds, certain weak acids, etc. are formed which interfere in the succeeding process by inhibiting the enzymes and fermenting microorganisms. However, this abundantly available form of sugar can neither be used as feed nor fodder, thus making it a potent source for bioenergy conversion. The different processes of pretreatment can be broadly classified as chemical, physicochemical, physical and biological methods (Table 12.2).

Proper pretreatment process increases the exposure of the holocellulose towards the saccharifying enzymes which lead to the conversion of polymerized sugar chains into their respective monomer units and thus, the efficacy of the entire process gets improved (Gujjala et al. 2016; Kumar et al. 2017c). In general, the majority of

Table 12.3 Lignocellulosics content of different biomass widely used for bioethanol production (Sun et al. 2016)

Lignocellulosic biomass	Examples	Holocellulose content		Lignin content (%)
		Cellulose (%)	Hemicellulose (%)	
Angiosperms	Oak, Maple. Walnut, Beech, etc.	40–55	24–40	18–25
Gymnosperms	Pine, Spruce, Cedar, Willow, etc.	45–50	25–35	25–35
Stalk	Wheat, Paddy, Corn, Rye, Barley, etc.	30–32	24–50	15–18
Baggase	Sugarcane, sorghum, etc.	45–55	20–25	18–24
Energy crops	Switch grass, Napier grass, *Miscanthus sp.* etc.	25–40	35–50	10–30

the lignocellulosics contain cellulose and hemicelluloses, followed by lignin, pectin and esters. Holocellulose being the chain polysaccharides are tightly associated with heterogenous cross-linked phenyl-propane lignin polymer (Banerjee et al. 2017) and is designated as the fundamental carbohydrate present in the plant cell wall (Althuri and Banerjee 2017; Sherpa et al. 2017). These tight associations via the covalent and non-covalent interactions form the intermeshed lignocellulosic matrix which results in low digestion rate and high recalcitrance to hydrolyzing enzymes (Rajak and Banerjee 2018). Enzymatic saccharification of holocellulose is another essential step in the conversion of lignocellulosics to biofuels.

Bioethanol from lignocellulosics is a promising technology involving pretreatment or delignification, hydrolysis or saccharifiaction, fermentation and distillation as its fundamental steps. Although, being a promising technology for meeting the global energy crisis, it has several bottlenecks which make the technology less popular. Some of the major concerns that lie among the people are lack of awareness, acceptability, ethical issues, etc. Presently, research on 2G ethanol production has taken place at various levels at its own pace where Government attention is required for funding the future research and create awareness among the industrialists as well as common men for its availability and acceptance.

4.2 Biobutanol

Globally butanol is produced from fossil fuel feedstocks. It is an important compound that have many applications such as in the production of solvents, plasticizers, butylamines, amino resins, butyl acetates and so on. Butanol is considered as a superior biofuel in comparison to ethanol due to its similar energy content like petroleum fuels and lower vapour pressure than ethanol (Zhang et al. 2014). Because of its low vapour pressure, butanol can be safely used in transportation motors. Butanol can be used as automobile fuel with little modifications of the engines. Moreover, the current gas station and transport infrastructures can keep on being utilized for butanol transport without modification as butanol is lesser hygroscopic and destructive to the pipelines than bioethanol (Calahan et al. 2015). At present, biobutanol is not utilized as a biofuel because of a few difficulties like the cost of feedstock, high toxicity of butanol to fermenting microbes and lastly, the low yield of butanol production, which results in consumption of high energy and hence, in the increment of the end product.

The differences between ethanol and butanol mentioning the fuel quality with various parameters has been represented in Table 12.4.

ABE Fermentation

The conversion of lignocellulosics to biobutanol, an enhanced liquid biofuel that can be used as transportation fuel, follows the ABE (Acetone-Butanol-Ethanol) fermentation pathway (Kujawska et al. 2015; Kumar et al. 2017c) as mentioned in Fig. 12.2. But due to the toxicity of butanol, ABE fermenting microbes cannot

Table 12.4 Fuel quality of ethanol and butanol (Kujawska et al. 2015)

Properties	Bioethanol	Biobutanol
Density at 20° (g/ml)	0.79	0.81
Vapour pressure at 20° (kPa)	7.58	0.53
Motor octane number	102	78
Boiling point (°)	78–78.4	117–118
Freezing point (°)	−114 to −114.5	−89.3 to −89.5
Air-fuel ratio	9.0	11.2
Flash point (°)	13	35
Autoignition temperature (°)	363	345
Energy density (MJ/L)	19.6	27–29.2

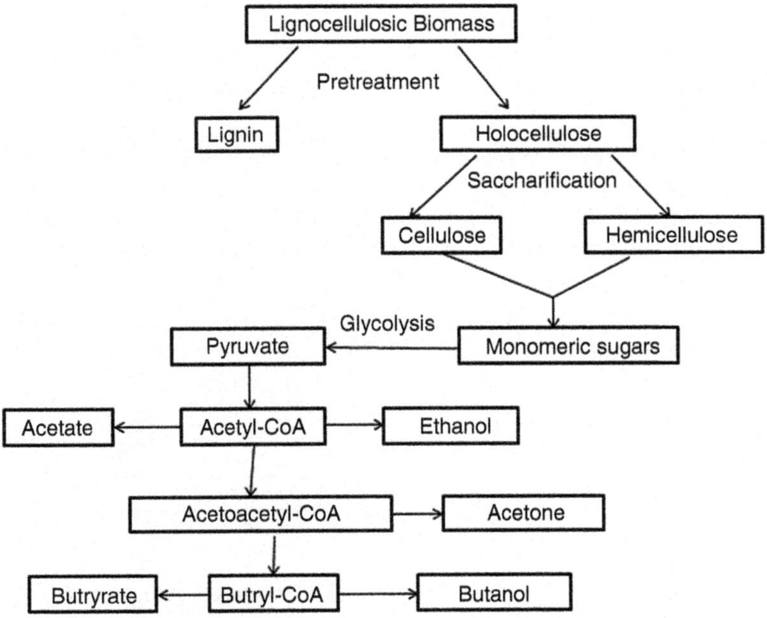

Fig. 12.2 Schematic diagram of ABE fermentation

tolerate acetone more than 20–30 g/L in batch fermenters, in which the order of butanol is 13–18 g/L (Qureshi et al. 2013). To outcast this problem several strategies have been employed such as developing high butanol tolerant strains using genetic manipulations, applying advance process designing to deal with the recovery of butanol from the fermentation process, and many more.

The motive behind the process configuration is to combine more than one operation into single unit to redeem the operational cost. Biobutanol production greatly depends on the factors like proper pretreatment of feedstock, fermentation and recovery of butanol at cost effective processes.

4.3 Biodiesel

Biodiesel is a renewable, biodegradable and eco-friendly biofuel that can be used as petro-diesel blends or even in pure form (Garlapati et al. 2013; Singh et al. 2012). It can be defined as fatty acid methyl ester (FAME) generated by transesterification of various lipid feedstock (Leung et al. 2010; Singh et al. 2012). These feedstocks can be obtained from plants, animals, microalgae, oleaginous micobes, etc. (Gujjala et al. 2017; Karmakar et al. 2010; Kumar et al. 2017b; Das et al. 2017). The production process includes major steps like the extraction of lipids, transesterification and FAME purification (Kumar et al. 2017b). Figure 12.3 shows the schematic representation of biodiesel production. In the subsequent section biodiesel from non-edible oil seed, animal fat, vegetable wastes oil has been described.

4.3.1 Non-edible Oil Seed, Animal Fat, Vegetable Wastes Oil

Biodiesel production has been successfully applied on the lipid extracted from plants such as *Pongamia pinnata*, *Simarouba glauca*, Jatropha, *Azadirachta indica*, etc. (Kumari et al. 2008, 2009; Garlapati et al. 2013; Verma and Sharma 2016) and from animal fats such as beef tallow, pork lard, and chicken fat (Bankovic-Ilic et al. 2014; Mata et al. 2014; Chakraborty et al. 2014).

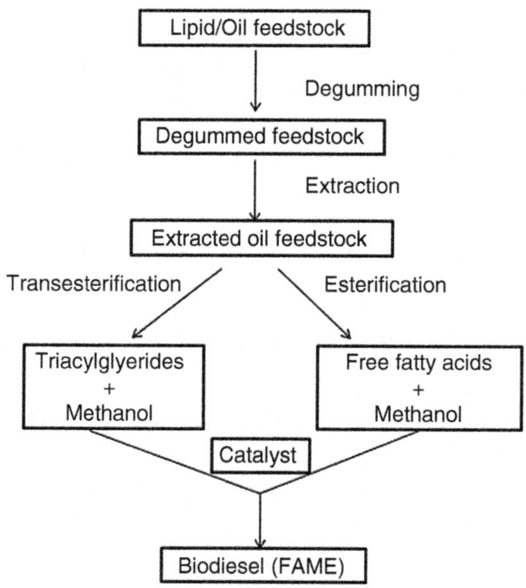

Fig. 12.3 Schematic representation of biodiesel production

Now-a-days, wastes trap grease generated from restaurants and households are getting popular towards biodiesel production (Kulkarni and Dalai 2006; Canakci 2007; Zhang et al. 2003). The amount of wastes trap grease generated varies from 1406 to 11,000 kg per year per restaurant (Hums et al. 2016) and the lipid content in wastes trap grease lies between 0.1% and 40% (Ward 2012). According to a study conducted by Ragauskas et al. (2013), the biodiesel production from grease interceptor is an efficient process and can be accomplished (Montefrio et al. 2010; Ragauskas et al. 2013). The major bottleneck in conversion of this wastes grease to biodiesel is free fatty acid (FFA) content (Lou et al. 2008; Sheng and Feng 2015). During alkali mediated transesterification, these FFA interfere by soap formation (Alsoudy and Janajreh 2012; Narwal and Gupta 2013; Atadashi et al. 2012). This issue can be overcome by enzymatic mediated transesterification where the enzymes carry out both esterification as well as transesterification (Gujjala et al. 2017; Kumar et al. 2017b; Fjerbaek et al. 2009).

4.3.2 Algal Biomass

The third generation biomass microalgae are considered as one of the most important substitution of oil source for biodiesel production (Dragone et al. 2010; Das et al. 2017; Nautiyal et al. 2014). These are one of the simplest photosynthetic microorganisms that can be easily grown on a simple culture medium (Sander and Murthy 2010; Milledge and Heaven 2013). Microalgae are becoming a promising supplement oil source for biodiesel production. It is easily grown on a simple culture medium. It can live in harsh condition due to their unicellular or simple multicellular structure (Kumar et al. 2017b; Mata et al. 2010; Varshney et al. 2015). Microalgae exhibit different types of metabolic pattern which includes photoautotrophic, heterotrophic, photoheterotrophic and myxotrophic (Mata et al. 2010; Tomaselli 2004; Barsanti and Gualtieri 2014).

Microalgae have been suggested as one of the most potential candidate for the biodiesel production as it has several advantages which include high photosynthetic efficiency, higher biomass production and faster growth rate as compared to other energy crops (Mata et al. 2010). However, the use of microalgal biodiesel is restricted due to its high production cost (Chisti 2007; Norsker et al. 2011; Slade and Bauen 2013). In order to solve this problem, there is a need to develop a production process to increase the biomass production and lipid content (Chisti 2007). It is well known that microalgae lipid yield is strain-dependent, but, it is clearly advantageous due to the higher biomass productivity and oil yield than those of other biofuel production (Liang et al. 2009; Li et al. 2014; Dash and Banerjee 2017).

5 Gaseous Biofuels

Apart from the solid and liquid biofuels, gaseous biofuels represent another important group of biofuel for energy production which is used as cooking fuel, transportation fuel, as well as in electricity generation (Vamvuka 2011; Stamatelatou et al. 2010; Kumar et al. 2017a). It can be stored in liquid compressed form which is economically advantageous in transportation process (Browne et al. 2011). Biogas, biohydrogen and biosyngas are few examples of gaseous biofuels (Kumar et al. 2017a; Izquierdo et al. 2012; Alves et al. 2013).

5.1 Biogas

A mixture of various gases mainly methane and carbon dioxide produced by anaerobic digestion of organic matter constitute the biogas. It can be produced from raw materials such as agricultural wastes, municipal wastes, plant material, food wastes, etc. The production involves anaerobic digestion with methanogens or anaerobic organisms (Kumar et al. 2017a; Izquierdo et al. 2013; Alves et al. 2013). In this process, the materials are digested inside a closed system called anaerobic digester, biodigester or a bioreactor (Ainsworth et al. 2003; Bouallagui et al. 2005). Upon combustion and oxidation in presence of oxygen, the energy released is used as a fuel for various purposes like cooking, heating, generating electricity, etc. (Vamvuka 2011; Ras et al. 2011). Biomethane could be formed from biogas by removal of some undesirable gases like carbon dioxide and high methane concentration (95–99%) could be achieved (Khan et al. 2017; Ryckebosch et al. 2011).

5.2 Biomethane

The biomethane production has several applications in generation of energy and can also be used as transportation fuel (Khan et al. 2017; Ryckebosch et al. 2011). The reuse of municipal solid wastes towards biomethane production can solve various global issues by converting the pollution causing wastes into energy sources (Browne et al. 2014). It will not only solve the wastes disposal and reduce the level of pollution but also act as an alternative energy source (Paepatung et al. 2009; Angelidaki et al. 2009; Singh et al. 2012). The production of biomethane involves three major steps viz., hydrolysis, acidogenesis followed by methanogenesis (DiMarco et al. 1990; Ryckebosch et al. 2011; Ferry 2012). These processes can either be chemical or microbial based (Singh et al. 2012; Ferry 2012). The post-harvest wastes is also used along with manure and microbial consortia for the production of biomethane (Kumar et al. 2017a). Sometimes, the solid wastes may contain lignocellulosic material and such types of wastes require pretreatment before its conversion to biomethane (Singh et al. 2012).

5.3 Biohydrogen

Biohydrogen, a clean renewable energy, can be produced from various types of organic wastes. Certain life forms such as hydrogen producing bacteria, algae, and archae are the main biological agents for biohydrogen production (Beer et al. 2009; Levin et al. 2004). Unlike burning fossil fuels, combustion of hydrogen releases meagre or no greenhouse gas into the environment.

Different groups of photosynthetic bacteria and algae have been studied extensively for hydrogen production where the light as well as dark reactions are playing a significant role. Hydrogen holds approximate 2.75 times higher energy density compared to other fuels. Biohydrogen is considered as future fuel due to its high conversion efficiency, recyclability, and nonpolluting nature. Although hydrogen can be produced from chemical as well as biological methods, the production of hydrogen through chemical routes are also getting importance because of its high productivity and low cost of production. Coal gasification, electrolysis of water, pyrolysis and biomass gasification have been extensively studied for hydrogen production through non-biological routes. Basically, there are four ways defined to produce biohydrogen biologically viz., (a) Direct/Indirect water photolysis, (b) Use of photosynthetic bacteria, (c) Fermentation by anaerobic microorganism, and, (d) Simultaneous dark and photo fermentation. Depending on the biochemical characteristics of the waste, like volatile solid content, moisture content, biodegradability, carbohydrate and C/N ratio, production yield has been varied (Kovács et al. 2006; Cheng and Logan 2007).

The major setback in biohydrogen production is proper conversion of organic material into end product. Furthermore, microorganisms associated in biohydrogen production have their own limitation and tolerance limit in high hydrogen concentration. Optimization of process pathway and engineered microorganism may facilitate higher biohydrogen yield. Both basic and applied R&D activities on biological production of hydrogen have drawn significant attention over past few decades. Research progress for identifying the major techno-economic bottlenecks of various bioprocesses for commercial production of hydrogen appears promising. However, high hydrogen yield remains to be the ultimate goal and challenge for the biohydrogen research and development. Enhancement in hydrogen yield may be possible by using suitable microbial strain, process modification, efficient bioreactor design and also genetic and metabolic engineering technique, to redirect metabolic pathway.

5.4 *Biosyngas*

Biosyngas (synthesis gas) contains mainly hydrogen and carbon monoxide produced due to partial combustion of biomass, and is used in the production of several industrial products like methanol, oils, diesel, ammonia, methane, Fischer-Tropsch liquids etc. Conventionally, syngas is produced from fossil fuels like coal, natural gas, etc. through a high temperature process using gasifiers operating at around 1600°. But biosyngas from renewable resources, such as biomass, exhibits a promising prospective. Panigrahi et al. (2003) explored synthesis gas production from steam gasification of biomass-derived oil (BDO). BDO is gasified at 800° under atmospheric pressure in a reactor using mixtures of CO_2 and N_2, H_2 and N_2, and steam. The biosyngas produced from this process comprised H_2, CO, CO_2, CH_4, C_2, C_3 and C_4^+. Chaudhari et al. (2003) investigated synthesis gas production from biomass-derived char (BDC). BDC was pyrolised in the presence of nitrogen at around 700–800° and then steam gasification at 650–800° was carried out in a tubular reactor to produce biosyngas having very high H_2/CO molar ratio (about 5–6). Wang et al. (2015) studied biosyngas from reforming of biomass gasification gas by addition of biogas. The biosyngas produced through this method has H_2/CO ratio above 1.5 and traces of CH_4 and CO_2. From the scientific studies, a huge potentiality of utilizing organic waste to syngas explores the future possibilities for appropriate utilization and use of syngas as one of the alternative energy sources (Lv et al. 2007; Wang et al. 2014).

6 Valorization of Wastes and Value-Added Products

The wide ranges of byproducts accumulated during pretreatment of biomass are the resultant product from modification of monomeric sugar moieties. Furfural, hydroxymethylfurfurals (HMF), levulinic acid, phenol, formic acid, acetic acid, etc. are some of the pretreatment byproducts with inhibitory effects in saccharification and fermentation of sugars (Zha et al. 2012).

HMF and levulinic acid are high value low products yield during pretreatment of lignocellulosics. These compounds can be formed from many different feedstocks including monosaccharides, disaccharides and polysaccharides. These are produced by the catalytic enolization of hexoses through different intermediate cyclic compounds like enediol, 3,4-dideoxyglucosulosene-3, humic acid, etc. The main technical challenge associated with HMF and levulinic acid production is the application of robust conditions and proper process optimization. HMF are precursor molecules for producing levulinic acid, 2,5-furandicarboxylic acid (FDCA) which has emerged as potential replacement for terephthalic acid (TPA) in polyester (PET) production. It has many applications in industries such as in pharmaceuticals as an anti-inflammatory, anti-allergic agent; polymer and plasticizers; resins and coatings. However, as fuel additive, it has potential to replace the current cetane improvers (Morone et al. 2015).

Xylitol has gained its immense importance due to plethora of applications it offers to the industries. Ranging from food industries in gums; tooth rehardening, and re-mineralization properties; capability of preventing otitis ear and upper respiratory infections, it has possibility of being used as a sweetener in syrups, tonics, and vitamin formulations. However, the foremost utilization is for the prevention of dental caries as it inhibits the growth of microorganisms responsible for tooth decay (Chen et al. 2010; Ur-Rehman et al. 2015). Although hexose utilization for fermentation is a common process but the leftover pentose are of another concern. Wild yeast cannot utilize hexose, but there are certain microorganisms which can utilize pentose to form xylitol, thus making it a high value end product. Xylitol is an intermediate formed during fermentation of xylose to ethanol. It is considered as a high value low yield byproduct. At present, industrial production of xylitol is achieved by conventional catalytic hydrogenation method where the pentose sugar are exposed to high temperature and pressure for polyol formation. The harsh operating conditions, extraction and purification processes have made the production of xylitol very expensive. But the biotechnological approach for the production of xylitol through pentose utilizing microbes and engineered microbes have been under rigorous studies, and thus, increase the production of xylitol at large scale in a more techno-economic manner.

The organic acids produced during the fermentation are high value high yield product and have many applications in the food, chemical, pharmaceutical, comestic and tannery industries. Several valuable organic acids like the butyric acid, acetic acid, citric acid, succinic acid, itaconic acid, lactic acid, levulinic acid, etc. are generated as byproducts at several stages of the bioconversion of lignocellulosics to biofuel. Microorganisms intake the sugars and metabolize them through biological pathways and produce these organic acids.

7 Challenges and Future Scope

The waste to wealth with bioenergy process comes with various challenges. Valorization of wastes is not confined to science but it is more like art where many branches of scientific inputs such as engineering aspects along with ethical issues and asthetic values are coming into the picture. Government of India has taken utmost care under the banner of Swachh Bharat mission, where the waste management can be regulated under various legislative bodies in order to protect the environment and human health.

The major challenges lie at the types of wastes, its composition, collection, segregation, pretreatment and efficient conversion to bioenergy. The degree of wastes composition varies from site to site. However, there is still a considerable part of undesired end-products that must be taken care of, and a more suitable solution needs to be found which is a continuous process. Depending upon the waste generation, collection and treatment optimization, the authors are proposing the decentralized waste treatment processes where not only the desirable products/byproducts will be developed but also the opening of huge employment opportunities will be there.

The main technical challenge associated with the conversion of waste to bioenergy is colliding with the acceptance, accessibility, and techno-economic feasibility of the different processes. The widely accepted chemical, physical and physico-chemical treatment processes that are currently being used for energy generation need further research work because of the release of certain intermediates/byproducts which have direct impact on global warming. Thus, the major thrust to this generation researcher is to develop the process/products which will have zero/no emission which will make the process eco-friendly, sustainable and green-based technology. With the generation of value added biochemical, the overall economy of the existing practices will be further improved.

References

Ainsworth, J. L., Atwood, D. and Rideout, T. (2003). U.S. Patent No. 6,569,332. Washington, DC: U.S. Patent and Trademark Office.

Ala'a Alsoudy, M.H.T. and Janajreh, I. (2012). Influence on process parameters in transesterification of vegetable and waste oil–a review. *Int. J. of Res. Rev. Appl. Sci*, **10(1):** 64.

Althuri, A. and Banerjee, R. (2017). Separate and simultaneous saccharification and fermentation of a pretreated mixture of lignocellulosic biomass for ethanol production. *Biofuels*, 1–12.

Alves, H.J., Junior, C.B., Niklevicz, R.R., Frigo, E.P., Frigo, M.S. and Coimbra-Araújo, C.H. (2013). Overview of hydrogen production technologies from biogas and the applications in fuel cells. *International Journal of Hydrogen Energy*, **38(13):** 5215–5225.

Angelidaki, I., Alves, M., Bolzonella, D., Borzacconi, L., Campos, J.L., Guwy, A.J. and Van Lier, J.B. (2009). Defining the biomethane potential (BMP) of solid organic wastes and energy crops: A proposed protocol for batch assays. *Water Science and Technology*, **59(5):** 927–934.

Ariunbaatar, J., Panico, A., Esposito, G., Pirozzi, F. and Lens, P.N. (2014). Pretreatment methods to enhance anaerobic digestion of organic solid waste. *Applied Energy*, **123:** 143–156.

Atadashi, I.M., Aroua, M.K., Aziz, A.A. and Sulaiman, N.M.N. (2012). Production of biodiesel using high free fatty acid feedstocks. *Renewable and Sustainable Energy Reviews*, **16(5):** 3275–3285.

Avanthi, A., Kumar, S., Sherpa, K.C. and Banerjee, R. (2017). Bioconversion of hemicelluloses of lignocellulosic biomass to ethanol: An attempt to utilize pentose sugars. *Biofuels*, **8(4):** 431–444.

Banerjee, R., Chintagunta, A.D. and Ray, S. (2017). A cleaner and eco-friendly bioprocess for enhancing reducing sugar production from pineapple leaf waste. *Journal of Cleaner Production*, **149:** 387–395.

Banković-Ilić, I.B., Stojković, I.J., Stamenković, O.S., Veljkovic, V.B. and Hung, Y.T. (2014). Waste animal fats as feedstocks for biodiesel production. *Renewable and Sustainable Energy Reviews*, **32:** 238–254.

Barsanti, L. and Gualtieri, P. (2014). Algae: Anatomy, Biochemistry, and Biotechnology. CRC Press.

Batayneh, M., Marie, I. and Asi, I. (2007). Use of selected waste materials in concrete mixes. *Waste Management*, **27(12):** 1870–1876.

Beer, L.L., Boyd, E.S., Peters, J.W. and Posewitz, M.C. (2009). Engineering algae for biohydrogen and biofuel production. *Current Opinion in Biotechnology*, **20(3):** 264–271.

Bouallagui, H., Touhami, Y., Cheikh, R.B. and Hamdi, M. (2005). Bioreactor performance in anaerobic digestion of fruit and vegetable wastes. *Process Biochemistry*, **40(3–4):** 989–995.
Browne, J.D., Allen, E. and Murphy, J.D. (2014). Assessing the variability in biomethane production from the organic fraction of municipal solid waste in batch and continuous operation. *Applied Energy*, **128:** 307–314.
Browne, J., Nizami, A.S., Thamsiriroj, T. and Murphy, J.D. (2011). Assessing the cost of biofuel production with increasing penetration of the transport fuel market: A case study of gaseous biomethane in Ireland. *Renewable and Sustainable Energy Reviews*, **15(9):** 4537–4547.
Calahan, D., Blersch, D. and Adey, W. (2015). Weeds in the algae garden—A source of biomass for the algae-to-biofuels program.
Canakci, M. (2007). The potential of restaurant waste lipids as biodiesel feedstocks. *Bioresource Technology*, **98(1):** 183–190.
Chakraborty, R., Gupta, A.K. and Chowdhury, R. (2014). Conversion of slaughterhouse and poultry farm animal fats and wastes to biodiesel: Parametric sensitivity and fuel quality assessment. *Renewable and Sustainable Energy Reviews*, **29:** 120–134.
Chaudhari, S.T., Dalai, A.K. and Bakhshi, N.N. (2003). Production of hydrogen and/or syngas (H_2+CO) via steam gasification of biomass-derived chars. *Energy & Fuels*, **17(4):** 1062–1067.
Chen, X., Jiang, Z.H., Chen, S. and Qin, W. (2010). Microbial and bioconversion production of D-xylitol and its detection and application. *International Journal of Biological Sciences*, **6(7):** 834.
Cheng, S. and Logan, B.E. (2007). Sustainable and efficient biohydrogen production via electrohydrogenesis. *Proceedings of the National Academy of Sciences*, **104(47):** 18871–18873.
Chintagunta, A.D., Jacob, S. and Banerjee, R. (2016). Integrated bioethanol and biomanure production from potato waste. *Waste Management*, **49:** 320–325.
Chisti, Y. (2007). Biodiesel from microalgae. *Biotechnology Advances*, **25(3):** 294–306.
Das, H. and Singh, S.K. (2004). Useful byproducts from cellulosic wastes of agriculture and food industry—A critical appraisal. *Critical Reviews in Food Science and Nutrition*, **44(2):** 77–89.
Das, M., Kundu, D., Singh, J., Rastogi, A. and Banerjee, R. (2017). Physiology and biochemistry of indigenous tribal liquor haria: A state of art. *Advances in Biotechnology and Microbiology*, **6:** 1–5.
Dash, A. and Banerjee, R. (2017). Enhanced biodiesel production through phyco-myco co-cultivation of Chlorella minutissima and Aspergillus awamori: An integrated approach. *Bioresource Technology*, **238:** 502–509.
Davis, G. and Song, J.H. (2006). Biodegradable packaging based on raw materials from crops and their impact on waste management. *Industrial Crops and Products*, **23(2):** 147–161.
Demirbas, A. (2011). Waste management, waste resource facilities and waste conversion processes. *Energy Conversion and Management*, **52(2):** 1280–1287.
DiMarco, A.A., Bobik, T.A. and Wolfe, R.S. (1990). Unusual coenzymes of methanogenesis. *Annual Review of Biochemistry*, **59(1):** 355–394.
Dincer, I. (2000). Renewable energy and sustainable development: A crucial review. *Renewable and Sustainable Energy Reviews*, **4(2):** 157–175.
Dragone, G., Fernandes, B.D., Vicente, A.A. and Teixeira, J.A. (2010). Third generation biofuels from microalgae. *Current Research, Technology and Education Topics in Applied Microbiology and Microbial Biotechnology*, **2:** 1355–1366.
Dutta, D. and Goel, S. (2017). Applications of Remote Sensing and GIS in Solid Waste Management—A Review. *In:* Advances in Solid and Hazardous Waste Management (pp. 133–151). Springer, Cham.
Ferry, J.G. (2012). Methanogenesis: Ecology, Physiology, Biochemistry & Genetics. Springer Science & Business Media.
Fjerbaek, L., Christensen, K.V. and Norddahl, B. (2009). A review of the current state of biodiesel production using enzymatic transesterification. *Biotechnology and Bioengineering*, **102(5):** 1298–1315.

Garlapati, V.K., Kant, R., Kumari, A., Mahapatra, P., Das, P. and Banerjee, R. (2013). Lipase mediated transesterification of Simarouba glauca oil: A new feedstock for biodiesel production. *Sustainable Chemical Processes*, **1(1):** 11.

Goel, S. (2008). Municipal solid waste management (MSWM) in India: A critical review. *J Environ Sci Eng*, **50(4):** 319–328.

Goel, S. (Ed.). (2017). Advances in Solid and Hazardous Waste Management. Springer.

Grossman, G.M. and Krueger, A.B. (1995). Economic growth and the environment. *The Quarterly Journal of Economics*, **110(2):** 353–377.

Gujjala, L.K., Bandyopadhyay, T.K. and Banerjee, R. (2016). Kinetic modelling of laccase mediated delignification of Lantana camara. *Bioresource Technology*, **212:** 47–54.

Gujjala, L.K., Kumar, S.J., Talukdar, B., Dash, A., Kumar, S., Sherpa, K.C. and Banerjee, R. (2017). Biodiesel from oleaginous microbes: Opportunities and challenges. *Biofuels*, 1–15.

Gupta, R. and Mohapatra, H. (2003). Microbial biomass: An economical alternative for removal of heavy metals from waste water.

Hums, M.E., Cairncross, R.A. and Spatari, S. (2016). Life-cycle assessment of biodiesel produced from grease trap waste. *Environmental Science & Technology*, **50(5):** 2718–2726.

Izquierdo, U., Barrio, V.L., Lago, N., Requies, J., Cambra, J.F., Güemez, M.B. and Arias, P.L. (2012). Biogas steam and oxidative reforming processes for synthesis gas and hydrogen production in conventional and microreactor reaction systems. *International Journal of Hydrogen Energy*, **37(18):** 13829–13842.

Izquierdo, U., Barrio, V.L., Requies, J., Cambra, J.F., Güemez, M.B. and Arias, P.L. (2013). Tri-reforming: A new biogas process for synthesis gas and hydrogen production. *International Journal of Hydrogen Energy*, **38(18):** 7623–7631.

Jacob, S. and Banerjee, R. (2016). Modeling and optimization of anaerobic codigestion of potato waste and aquatic weed by response surface methodology and artificial neural network coupled genetic algorithm. *Bioresource Technology*, **214:** 386–395.

Kandakatla, P., Mahto, B. and Goel, S. (2013). Extent and rate of biodegradation of different organic components in municipal solid waste. *International Journal of Environment and Waste Management*, **11(4):** 350–364.

Karmakar, A., Karmakar, S. and Mukherjee, S. (2010). Properties of various plants and animals feedstocks for biodiesel production. *Bioresource Technology*, **101(19):** 7201–7210.

Kaza, S., Yao, L., Bhada-Tata, P. and Van Woerden, F. (2018). What a Waste 2.0: A Global Snapshot of Solid Waste Management to 2050. World Bank Publications.

Khan, I.U., Othman, M.H.D., Hashim, H., Matsuura, T., Ismail, A.F., Rezaei-Dasht Arzhandi, M. and Azelee, I.W. (2017). Biogas as a renewable energy fuel—A review of biogas upgrading, utilisation and storage. *Energy Conversion and Management*, **150:** 277–294.

Kovács, K.L., Maróti, G. and Rákhely, G. (2006). A novel approach for biohydrogen production. *International Journal of Hydrogen Energy*, **31(11):** 1460–1468.

Kujawska, A., Kujawski, J., Bryjak, M. and Kujawski, W. (2015). ABE fermentation products recovery methods—A review. *Renewable and Sustainable Energy Reviews*, **48:** 648–661.

Kulkarni, M.G. and Dalai, A.K. (2006). Waste cooking oil—An economical source for biodiesel: A review. *Industrial & Engineering Chemistry Research*, **45(9):** 2901–2913.

Kumar, K.N. and Goel, S. (2009). Characterization of municipal solid waste (MSW) and a proposed management plan for Kharagpur, West Bengal, India. *Resources, Conservation and Recycling*, **53(3):** 166–174.

Kumar, M., Upadrasta, L. and Banerjee, R. (2017a). Biomethanation of pineapple wastes using potent anaerobic consortia substituting cow manure. *Environmental Engineering & Management Journal (EEMJ)*, **16(11):** 2647–2655.

Kumar, P., Barrett, D.M., Delwiche, M.J. and Stroeve, P. (2009). Methods for pretreatment of lignocellulosic biomass for efficient hydrolysis and biofuel production. *Industrial & Engineering Chemistry Research*, **48(8):** 3713–3729.

Kumar, S.J., Kumar, G.V., Dash, A., Scholz, P. and Banerjee, R. (2017b). Sustainable green solvents and techniques for lipid extraction from microalgae: A review. *Algal Research*, **21:** 138–147.

Kumar, S., Gujjala, L.K. and Banerjee, R. (2017c). Simultaneous pretreatment and saccharification of bamboo for biobutanol production. *Industrial Crops and Products*, **101:** 21–28.

Kumari, A., Mahapatra, P., Garlapati, V.K. and Banerjee, R. (2009). Enzymatic transesterification of Jatropha oil. *Biotechnology for Biofuels*, **2(1):** 1.

Kumari, A., Mahapatra, P., Kumar, G.V. and Banerjee, R. (2008). Comparative study of thermostabilty and ester synthesis ability of free and immobilized lipases on cross linked silica gel. *Bioprocess and Biosystems Engineering*, **31(4):** 291–298.

Kundu, D., Dutta, D., Mondal, S., Haque, S., Bhakta, J.N. and Jana, B.B. (2017). Application of Potential Biological Agents in Green Bioremediation Technology: Case Studies. *In:* Handbook of Research on Inventive Bioremediation Techniques (pp. 300–323). IGI Global.

Leung, D.Y., Wu, X. and Leung, M.K.H. (2010). A review on biodiesel production using catalyzed transesterification. *Applied Energy*, **87(4):** 1083–1095.

Levin, D.B., Pitt, L. and Love, M. (2004). Biohydrogen production: Prospects and limitations to practical application. *International Journal of Hydrogen Energy*, **29(2):** 173–185.

Li, T., Zheng, Y., Yu, L. and Chen, S. (2014). Mixotrophic cultivation of a Chlorella sorokiniana strain for enhanced biomass and lipid production. *Biomass and Bioenergy*, **66:** 204–213.

Liang, Y., Sarkany, N. and Cui, Y. (2009). Biomass and lipid productivities of Chlorella vulgaris under autotrophic, heterotrophic and mixotrophic growth conditions. *Biotechnology Letters*, **31 (7):** 1043–1049.

Lou, W.Y., Zong, M.H. and Duan, Z.Q. (2008). Efficient production of biodiesel from high free fatty acid-containing waste oils using various carbohydrate-derived solid acid catalysts. *Bioresource Technology*, **99(18):** 8752–8758.

Lv, P., Yuan, Z., Wu, C., Ma, L., Chen, Y. and Tsubaki, N. (2007). Bio-syngas production from biomass catalytic gasification. *Energy Conversion and Management*, **48(4):** 1132–1139.

Mata, T.M., Martins, A.A. and Caetano, N.S. (2010). Microalgae for biodiesel production and other applications: A review. *Renewable and Sustainable Energy Reviews*, **14(1):** 217–232.

Mata, T.M., Mendes, A.M., Caetano, N.S. and Martins, A.A. (2014). Properties and sustainability of biodiesel from animal fats and fish oil. *Chemical Engineering Transactions*, **38:** 175–180.

Milledge, J.J. and Heaven, S. (2013). A review of the harvesting of micro-algae for biofuel production. *Reviews in Environmental Science and Bio/Technology*, **12(2):** 165–178.

Montefrio, M.J., Xinwen, T. and Obbard, J.P. (2010). Recovery and pre-treatment of fats, oil and grease from grease interceptors for biodiesel production. *Applied Energy*, **87(10):** 3155–3161.

Mor, S., Ravindra, K., De Visscher, A., Dahiya, R.P. and Chandra, A. (2006). Municipal solid waste characterization and its assessment for potential methane generation: A case study. *Science of the Total Environment*, **371(1–3):** 1–10.

Morone, A., Apte, M. and Pandey, R.A. (2015). Levulinic acid production from renewable waste resources: Bottlenecks, potential remedies, advancements and applications. *Renewable and Sustainable Energy Reviews*, **51:** 548–565.

Mosier, N., Wyman, C., Dale, B., Elander, R., Lee, Y.Y., Holtzapple, M. and Ladisch, M. (2005). Features of promising technologies for pretreatment of lignocellulosic biomass. *Bioresource Technology*, **96(6):** 673–686.

Narwal, S.K. and Gupta, R. (2013). Biodiesel production by transesterification using immobilized lipase. *Biotechnology Letters*, **35(4):** 479–490.

Nautiyal, P., Subramanian, K.A. and Dastidar, M.G. (2014). Production and characterization of biodiesel from algae. *Fuel Processing Technology*, **120:** 79–88.

Norsker, N.H., Barbosa, M.J., Vermuë, M.H. and Wijffels, R.H. (2011). Microalgal production—A close look at the economics. *Biotechnology Advances*, **29(1):** 24–27.

Olmstead, K.P. and Weber Jr, W.J. (1991). Interactions between microorganisms and activated carbon in water and waste treatment operations. *Chemical Engineering Communications*, **108 (1):** 113–125.

Paepatung, N., Nopharatana, A. and Songkasiri, W. (2009). Bio-methane potential of biological solid materials and agricultural wastes. *Asian Journal on Energy and Environment*, **10(1):** 19–27.

Panigrahi, S., Dalai, A.K., Chaudhari, S.T. and Bakhshi, N.N. (2003). Synthesis gas production from steam gasification of biomass-derived oil. *Energy & Fuels*, **17(3):** 637–642.

Qureshi, N., Liu, S. and Ezeji, T.C. (2013). Cellulosic butanol production from agricultural biomass and residues: Recent advances in technology. *In:* Advanced Biofuels and Bioproducts (pp. 247–265). Springer, New York, NY.

Ragauskas, A.M.E., Pu, Y. and Ragauskas, A.J. (2013). Biodiesel from grease interceptor to gas tank. *Energy Science & Engineering*, **1(1):** 42–52.

Rajak, R.C. and Banerjee, R. (2018). An eco-friendly process integration for second generation bioethanol production from laccase delignified Kans grass. *Energy Conversion and Management*, **157:** 364–371.

Ras, M., Lardon, L., Bruno, S., Bernet, N. and Steyer, J. P. (2011). Experimental study on a coupled process of production and anaerobic digestion of Chlorella vulgaris. *Bioresource Technology*, **102(1):** 200–206.

Ryckebosch, E., Drouillon, M. and Vervaeren, H. (2011). Techniques for transformation of biogas to biomethane. *Biomass and Bioenergy*, **35(5):** 1633–1645.

Sander, K. and Murthy, G.S. (2010). Life cycle analysis of algae biodiesel. *The International Journal of Life Cycle Assessment*, **15(7):** 704–714.

Sheng, J. and Feng, X. (2015). Metabolic engineering of yeast to produce fatty acid-derived biofuels: Bottlenecks and solutions. *Frontiers in Microbiology*, **6:** 554.

Sherpa, K.C., Ghangrekar, M.M. and Banerjee, R. (2017). Optimization of saccharifi-cation of enzymatically pretreated sugarcane tops by response surface methodology for ethanol production. *Biofuels*, 1–8. DOI: https://doi.org/10.1080/17597269.2017.1409058.

Singh, A., Kuila, A., Adak, S., Bishai, M. and Banerjee, R. (2011). Use of fermentation technology on vegetable residues for value added product development: A concept of zero waste utilization. *International Journal of Food and Fermentation Technology*, **1(2):** 173–184.

Singh, A., Kuila, A., Adak, S., Bishai, M. and Banerjee, R. (2012). Utilization of vegetable wastes for bioenergy generation. *Agricultural Research*, **1(3):** 213–222.

Singh, O.V. and Chandel, A.K. (2018). Sustainable Biotechnology—Enzymatic Resources of Renewable Energy. Springer, Switzerland.

Slade, R. and Bauen, A. (2013). Micro-algae cultivation for biofuels: Cost, energy balance, environmental impacts and future prospects. *Biomass and Bioenergy*, **53:** 29–38.

Smith, K.R. and Lee, Y.S. (1993). Urbanization and the environmental risk transition.

Srivastava, S.K. (2007). Green supply-chain management: A state-of-the-art literature review. *International Journal of Management Reviews*, **9(1):** 53–80.

Stamatelatou, K., Antonopoulou, G., Tremouli, A. and Lyberatos, G. (2010). Production of gaseous biofuels and electricity from cheese whey. *Industrial & Engineering Chemistry Research*, **50 (2):** 639–644.

Sun, S., Sun, S., Cao, X., and Sun, R. (2016). The role of pretreatment in improving the enzymatic hydrolysis of lignocellulosic materials. *Bioresource Technology*, **199:** 49–58.

Tomaselli, L. (2004). The microalgal cell. *Handbook of Microalgal Culture: Biotechnology and Applied Phycology*, **1:** 3–19.

Ur-Rehman, S., Mushtaq, Z., Zahoor, T., Jamil, A. and Murtaza, M.A. (2015). Xylitol: A review on bioproduction, application, health benefits, and related safety issues. *Critical Reviews in Food Science and Nutrition*, **55(11):** 1514–1528.

Vamvuka, D. (2011). Bio-oil, solid and gaseous biofuels from biomass pyrolysis processes—An overview. *International Journal of Energy Research*, **35(10):** 835–862.

Varshney, P., Mikulic, P., Vonshak, A., Beardall, J. and Wangikar, P.P. (2015). Extremophilic micro-algae and their potential contribution in biotechnology. *Bioresource Technology*, **184:** 363–372.

Verma, P. and Sharma, M.P. (2016). Review of process parameters for biodiesel production from different feedstocks. *Renewable and Sustainable Energy Reviews*, **62:** 1063–1071.

Walton, S.V., Handfield, R.B. and Melnyk, S.A. (1998). The green supply chain: Integrating suppliers into environmental management processes. *International Journal of Purchasing and Materials Management*, **34(1):** 2–11.

Wang, S., Wang, H., Yin, Q., Zhu, L. and Yin, S. (2014). Methanation of bio-syngas over a biochar supported catalyst. *New Journal of Chemistry*, **38(9):** 4471–4477.

Wang, Y.G., Sun, J.L., Zhang, H.Y., Chen, Z.D., Lin, X.C., Zhang, S., ... Fan, M.H. (2015). In situ catalyzing gas conversion using char as a catalyst/support during brown coal gasification. *Energy & Fuels*, **29(3):** 1590–1596.

Ward, P.M.L. (2012). Brown and black grease suitability for incorporation into feeds and suitability for biofuels. *Journal of Food Protection*, **75(4):** 731–737.

Wyman, C.E., Dale, B.E., Elander, R.T., Holtzapple, M., Ladisch, M.R., Lee, Y.Y., ... Saddler, J.N. (2009). Comparative sugar recovery and fermentation data following pretreatment of poplar wood by leading technologies. *Biotechnology Progress*, **25(2):** 333–339.

Zha, Y., Muilwijk, B., Coulier, L. and Punt, P.J. (2012). Inhibitory compounds in lignocellulosic biomass hydrolysates during hydrolysate fermentation processes. *J. Bioprocess Biotechniq*, **2 (1):** 112–122.

Zhang, Y., Dube, M.A., McLean, D.D.L. and Kates, M. (2003). Biodiesel production from waste cooking oil: 1. Process design and technological assessment. *Bioresource Technology*, **89(1):** 1–16.

Zhang, Z., Wang, T., Jia, M., Wei, Q., Meng, X. and Shu, G. (2014). Combustion and particle number emissions of a direct injection spark ignition engine operating on ethanol/gasoline and n-butanol/gasoline blends with exhaust gas recirculation. *Fuel*, **130:** 177–188.

Zheng, Y., Yanful, E.K. and Bassi, A.S. (2005). A review of plastic waste biodegradation. *Critical Reviews in Biotechnology*, **25(4):** 243–250.

Index

D. Sengupta et al. (eds.), *Treatment and Disposal of Solid and Hazardous Wastes*,
https://doi.org/10.1007/978-3-030-29643-8

The manufacturer's authorised representative in the EU is Springer Nature Customer Service Centre GmbH, Europaplatz 3, 69115 Heidelberg, Germany. If you have any concerns regarding our products, please contact ProductSafety@springernature.com

Printed and bound by CPI Group (UK) Ltd, Croydon, CR0 4YY
30/09/2026
02225491-0003